ACTA PHYSICA AUSTRIACA / SUPPLEMENTUM IV

SPECIAL PROBLEMS IN HIGH ENERGY PHYSICS

PROCEEDINGS OF THE
VI. INTERNATIONALE UNIVERSITÄTSWOCHEN
FÜR KERNPHYSIK 1967 DER KARL-FRANZENS-UNIVERSITÄT
GRAZ, AT SCHLADMING (STEIERMARK, AUSTRIA)
26th FEBRUARY–9th MARCH 1967

SPONSORED BY
BUNDESMINISTERIUM FÜR UNTERRICHT
BUNDESMINISTERIUM FÜR HANDEL, GEWERBE UND INDUSTRIE
THE INTERNATIONAL ATOMIC ENERGY AGENCY
STEIERMÄRKISCHE LANDESREGIERUNG AND
KAMMER DER GEWERBLICHEN WIRTSCHAFT FÜR STEIERMARK

EDITED BY
PAUL URBAN
GRAZ

WITH 33 FIGURES

1967

SPRINGER-VERLAG WIEN GMBH

Organizing Committee:

Chairman: Prof. Dr. PAUL URBAN
Vorstand des Institutes für Theoretische Physik, Universität Graz

Committee Members: Dr. R. BAIER
Dr. P. BREITENLOHNER

Secretary: A. SCHMALDIENST

Acta Physica Austriaca / Supplementum I
Weak Interactions and Higher Symmetries
published in 1964

Acta Physica Austriaca / Supplementum II
Quantum Electrodynamics
published in 1965

Acta Physica Austriaca / Supplementum III
Elementary Particle Theories
published in 1966

Library of Congress Catalog Card Number: 67-26417

ISBN 978-3-211-80836-8 ISBN 978-3-7091-5485-4 (eBook)
DOI 10.1007/978-3-7091-5485-4
Softcover reprint of the hardcover 1st edition 1967

Titel-Nr. 9222

Preface

The past meetings held under the auspices of the Austrian Ministry of Education and the successes and benefits derived from these lectures in former years have encouraged us to continue this year with a survey of high energy physics.

The experimental side of this field has undergone such a great advancement that the evaluation of huge masses of data called for the theoretical physicist to develop new theories and to explain facts and connections which we have not been able to clarify up to the present. In spite of the precarious situation especially in the field of Elementary Particle Theories, the theoretical physicist is about to summarize the present situation and to give further impulses for the progress of our knowledge in nuclear physics. Thus the results of the Sixth International Meeting in Schladming and the interest shown in it fully justify our endeavours in this direction. In organizing the publication of the proceedings we shall continue on the same lines as hitherto so as to reduce delays through editing to an absolute minimum and to offer the publication at a reasonable price. I am again grateful for the assistance of the Springer-Verlag whose photo-mechanical method proved so successful. Nevertheless we wish to apologize for any mistakes that might have occurred in writing the text of formulae.

I am much indebted to my staff for their help and assistance in organizing our meeting, and I wish to take this opportunity to express my sincere appreciation to all those who unselfishly helped to make this meeting a success.

Graz, May 1967 **P. Urban**

Contents

Dear Colleagues, Ladies and Gentlemen:

In addressing you today upon the opening of the sixth International Meeting at Schladming, I want to express my satisfaction that we are again here together continuing the good work of five previous seminars. The fact that we assemble today is proof enough of the popularity of our efforts and gives me sufficient encouragement to continue on the same lines and we all hope that also the 6th International Meeting to which I welcome you most cordially will bring the benefit of our work to you and the countries you represent.

It is a great privilege for me to see among our guests of honour so many old friends who have given such important contributions and inspiration to our seminar. It is most gratifying to welcome so many distinguished members of our Federal Government and Provincial authorities giving proof of their serious interest in our work.

It is a great honour for me to welcome Herrn Ministerialrat Dr. Walter Hafner as representative of the Minister of Education, to whom we are indebted for sponsorship and grants which are the basis of our undertaking. Such grants are also received from the International Atomic Energy Agency as well as from the Ministry of Trade and Industry. We are further indebted for support to the Provincial Government whose representative Hofrat Dr. Pulitzky I herewith also welcome most heartily as well as Dr. Töpfner from the Chamber of Commerce.

It is a special pleasure for me to have among our guests of honour the Chancelor of the University of Graz, Prof. Dr. Möse, the representative of the Chancelor of the Technische Hochschule of Graz, Prof. Dr. Ledinegg, and the Director of Österreichische Studiengesellschaft für Atomenergie, Prof. Dr. Higatsberger. Moreover I want to express my special thanks to all those who have participated and I am confident that our present efforts will be as successfull as in the former years.

I trust you will enjoy staying here with us again in the beautiful mountains of Styria, as a setting for our Meeting, since the surroundings are certainly contributing towards the success of our Symposium. When speaking of our host city Schladming, we must not forget its representative, the Mayor of the City, Director Laurich, to whom we owe so much.

I will, no doubt, interest my listeners to hear that the number of participants of this year's meeting again is almost 170 representing as many as 12 nations. It is a great honour for me to have succeeded in gaining the cooperation of highly outstanding lecturers from abroad, and I want to express my gratitude that they undertook the trouble of coming to us, which will be rewarded by the benefit their knowledge will give us.

Now let me say a few words about this year's scientific program: We try to give a detailed survey of high energy physics which today attracts such world wide interest as a basis of every further nuclear physical research. While the description of the motion of electrons inside the atom was given with the aid of the quantum theory, a new branch of physics, nuclear physics, found two new types of interaction, the so-called strong interaction, which binds the nucleons in the nuclei, and the weak interaction, which is responsible for the ß-radioactivity. The study of these forces leads us beyond the field of nuclear physics so far known, for the research of such physical phenomena requires the investigation of matter at sub-nuclear dimensions. Now we know that the smaller the object we want to study, the higher the energy which is necessary to penetrate it. In recent times we succeeded in constructing accelerators of ever increasing energy, so that we have in our hands the necessary tools. With this requisit it was possible to discover a great number of new particles and also their properties. This new branch of physics is now identical with high-energy physics and the progress in this field is very significant for our knowledge of nature.

I trust also this year's symposium will bring us a step further in our understanding of

"High Energy Physics"

and in opening this meeting I wish it every success.

P. Urban, Graz

MESON DECAYS†

By

Laurie M. BROWN
Northwestern University, Evanston, Illinois

The following lectures, pedagogical in nature, introduce and illustrate with simple examples, some techniques and ideas which are useful in the phenomenological analysis of meson decay. Generally the form factors (or vertex functions) which contain most of the dynamics will be little discussed; the emphasis will be on kinematical aspects.

I. General Quantum Mechanical Framework*

1. S- and T-matrices

This is a sketch of the essentials:
Scattering and decay processes are described as transitions between states of non-interacting particles. Thus:

$$|\alpha \text{ in}\rangle \rightarrow |\beta \text{ out}\rangle \tag{1}$$

† Lecture given at the VI. Internationalen Universitätswochen f. Kernphysik, Schladming, 26 February-11 March 1967.
* see reference 1.

One particle states, either "in" or "out", are specified by three-momentum $\vec{k}$ and spin component s, thus $|\vec{k},s\rangle$. If the spin quantization direction is chosen as $\vec{k}$, then s is the helicity component. The one-particle states are normalized

$$\langle\vec{k}',s'|\vec{k},s\rangle = \delta_{ss'}(2\pi)^3\delta(\vec{k}'-\vec{k}) \tag{2}$$

and form a complete set

$$\Sigma_{s'}\int|\vec{k}',s'\rangle\langle\vec{k}',s'|\vec{k},s\rangle \frac{d\vec{k}'}{(2\pi)^3} = |\vec{k},s\rangle \quad . \tag{3}$$

Many particle "in" and"out" states are products of one-particle states.

To the transition (1) corresponds the transition amplitude

$$\langle\beta \text{ out} \mid \alpha \text{ in}\rangle \tag{4}$$

The "in" states are connected to the "out" states by a unitary transformation $S(S\,S^+ = S^+\,S = 1)$, so that

$$|\alpha \text{ in}\rangle = S|\alpha \text{ out}\rangle \tag{5}$$

and the transition amplitude can be written

$$\langle\beta \text{ out} \mid \alpha \text{ in}\rangle = \langle\beta \text{ out} |S|\alpha \text{ out}\rangle \equiv \delta_{\beta\alpha} \tag{6}$$

Or we can write

$$\delta_{\beta\alpha} = \langle\beta \text{ out}|(S^+S)S|\alpha \text{ out}\rangle =$$

$$= (\langle\beta \text{ out}|S^+)S(S|\alpha \text{ out}\rangle) = \langle\beta \text{ in}|S|\alpha \text{ in}\rangle \quad . \tag{7}$$

Since a part of S does not induce transitions, it is conventional to define T by

$$S = 1 + iT \tag{8}$$

and to write the matrix element of T between initial (i) and final (f) states as

$$T_{fi} = (2\pi)^4 \, \delta^4 \, P_f - P_i) \, \frac{F_{fi}}{[\Pi_j (2E_j)]^{1/2}} \tag{9}$$

The product Π_j is over all the particles belonging to the state i <u>and</u> all the particles belonging to the state f, E_j being the relativistic energy of the j-th particle.

F_{fi}, defined by (9), is called the <u>invariant</u> matrix element and is the same as the Feynman amplitude, calculated according to the rules given by Feynman, except that a factor $\sqrt{2m}$ has been included for each initial and final fermion, m being the fermion mass.

2. Cross sections and decay rates

Transition rates are obtained by summing the absolute square of the transition amplitude over the appropriate final states f; schematically we write

$$\Gamma = \Sigma_f |T_{fi}|^2 \quad . \tag{10}$$

In performing this sum (as discussed in Refs. 1 and 2), one factor of $(2\pi)^4 \, \delta^4(P_f - P_i)$ is replaced by

$$\Pi_{\text{fin}\, j} \frac{d\vec{k}_j}{(2\pi)^3} \tag{11}$$

where j labels the particles belonging to the final state f. In these lectures we are concerned with decays, but for completeness we define here the collision cross section for two particles:

$$\sigma = \frac{1}{\text{flux}} \Sigma_f |T_{fi}|^2 =$$

$$= \frac{1}{\text{flux}} \Pi_{\substack{i \\ inc}} (2E_i)^{-1} \int |F_{fi}|^2 (2\pi)^4 \delta^4 (P_f - P_i) \times$$

$$\times \Pi_{\substack{j \\ fin}} \left[\frac{d\vec{k}}{2E(2\pi)^3}\right]_j \tag{12}$$

For the collision of two particles of masses m_i and four-momenta q_i, this can be written in covariant form using

$$[\Pi_{inc} (2E_i)] \cdot \text{flux} = 4[(q_1 \cdot q_2)^2 - (m_1 . m_2)^2]^{1/2} \tag{13}$$

and

$$\frac{1}{2E} = \int \delta(k^2 - m^2) dk^o \; \Theta(k^o) \tag{14}$$

$$\Theta(x) = 1 \qquad x > 0$$

$$= 0 \qquad x < 0$$

The covariant form of the cross section is then

$$\sigma = \frac{1}{4[(q_1 q_2)^2 - (m_1 m_2)^2]^{1/2}} \int |F_{fi}|^2 (2\pi)^4 \times$$

$$\times \delta^4 (q_1 + q_2 - \Sigma_{\substack{j \\ fin}} k_j) \; \Pi_{\substack{j \\ fin}} \Theta(k_j^o) \delta(k_j^2 - m_j^2) \frac{d^4 k_j}{(2\pi)^3} \,. \tag{15}$$

Similarly, the total decay rate of a particle, obtained by summing (10) over all possible final states f is

$$\Gamma = \Sigma_f |T_{fi}|^2$$

$$= \text{probability of decay per second}$$
$$= \text{inverse lifetime} = \frac{1}{\tau} \tag{16}$$

The partial decay rate to a specified final state a is

$$\Gamma_a = (2M)^{-1}\int |F^a_{fi}|^2 (2\pi)^4 \delta^4(P_f - P_i) \prod_{fin\,j} \left[\frac{d\vec{k}}{2E(2\pi)^3}\right]_j . \quad (17)$$

M is the mass of the decaying particle assumed at rest and j denotes a particle of the final state a. (17) can obviously be put into a manifestly covariant form analogous to (15).

3. Phase space

The factors appearing under the integral sign in Eqs. (12), (15) and (17) make up a differential element of invariant phase space. This is the ordinary phase space element divided by $\prod_{fin\,j}(2E_j)$. It is useful to carry out some of the integrations for the 2- and 3-body cases:

a) 2-body invariant phase space

We consider the decay of a particle of 4-momentum q into two particles of 4-momentum $p_\pm = (E_\pm, \vec{p}_\pm)$. We have thus $q^2 = M^2 \equiv s$ and $p_\pm^2 = \mu_\pm^2$.

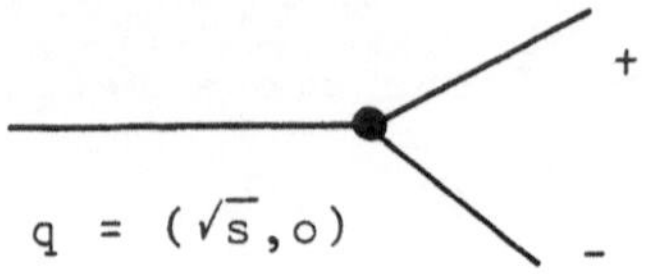

$q = (\sqrt{s}, o)$

Fig. 1

For the decaying particle at rest, $q = (\sqrt{s}, o)$. The two-body invariant phase space is

$$d\rho_2 = (2\pi)^{-6}\int d^4p_+ d^4p_- \delta(p_+^2 - \mu_+^2)\delta(p_-^2 - \mu_-^2)\Theta(p_+^o)\Theta(p_-^o) \times$$

$$\times (2\pi)^4 \delta^4(p_+ + p_- - q) \quad . \quad (18)$$

Integrate over d^4p_- and dp_+^o to get

$$d\rho_2 = (2\pi)^{-2}\int(2E_+)^{-1}d^3p_+\delta([q - p_+]^2 - \mu_-^2) \quad . \tag{19}$$

The delta function is

$$\delta(s + \mu_+^2 - \mu_-^2 - 2\sqrt{s}E_+) = \frac{1}{2\sqrt{s}}\,\delta(E_+ - \frac{s+\mu_+^2-\mu_-^2}{2\sqrt{s}}) \tag{20}$$

and

$$d^3p_+ = p_+E_+\,dE_+\,d\Omega_+ \tag{21}$$

so that

$$d\rho_2 = \frac{p_+}{4\pi\sqrt{s}}\;\frac{d\Omega_+}{4\pi} \tag{22}$$

with

$$p_+ = \left[\frac{(s + \mu_+^2 - \mu_-^2)^2 - 4s\mu_+^2}{4s}\right]^{1/2} \tag{22a}$$

b) 3-body invariant phase space

Similarly to the above, we consider the decay of a particle labelled η into three particles labelled +, -, and o. Without making any dynamical assumption we formally consider the + and - particles to make a quasi-particle of effective mass $\sqrt{s}$, i.e., we <u>define</u>

$$s \equiv (p_+ + p_-)^2 = (p_\eta - p_o)^2 \ . \tag{23}$$

Then, by inspection,

$$d\rho_3 = (2\pi)^{-3}\int d^4p_o\delta(p_o^2 - \mu_o^2)\Theta(p_o^o)d\rho_2(s) \ \cdot \tag{24}$$

Since

$$s = m_\eta^2 + \mu_o^2 - 2m_\eta\,E_o$$

we have

$$p_o = \left[\frac{(m_\eta^2 + \mu_o^2 - s)^2}{(2m_\eta)^2} - \mu_o^2\right]^{1/2} \tag{25}$$

and

$$d\rho_3 = (2\pi)^{-2}\int p_o dE_o \rho_2(s) = (2\pi)^{-2}(2m_\eta)^{-1}\int p_o(s)ds\rho_2(s)$$

$$= (32\pi^3)^{-1}(2m_\eta)^{-2}\int\frac{ds}{s}[(s+\mu_+^2-\mu_-^2)^2 - 4s\mu_+^2]^{1/2} \times$$

$$\times\ [(s - \mu_o^2 - m_\eta^2)^2 - 4m_\eta^2\mu_o^2]^{1/2} \quad . \tag{26}$$

We have here also integrated $d\Omega_+/4\pi$ to 1 .
Note that identity

$$(a + b + c)^2 - 4bc = (a + b - c)^2 + 4ac$$

can be used to write the last factor of (26) as

$$[(s + m_\eta^2 - \mu_o^2)^2 - 4sm_\eta^2]^{1/2} \quad . \tag{27}$$

In fact, let

$$\phi(a,b,c) \equiv [(a + b - c)^2 - 4ab]^{1/2} \tag{28}$$

Then

$$d\rho_2 \equiv \frac{1}{8\pi s}\,\phi(s,\ \mu_+^2,\ \mu_-^2)\ \frac{d\Omega_+}{4\pi} \tag{29}$$

and

$$d\rho_3 = \frac{1}{32\pi^3}\,\frac{1}{(2m_\eta)^2}\int\frac{ds}{s}\,\phi(s,\mu_+^2,\mu_-^2)\phi(s,m_\eta^2,\mu_o^2) \quad . \tag{30}$$

Similar iterative forms can be developed for more particles in the final state.

II. Two-Body Decay

Taking the decaying particle of mass M to be at rest (refer to Fig. 1), the decay angular distribution is given by

$$d\Gamma = (2M)^{-1} \sum_{\text{Spins}} |F_{fi}|^2 \, d\rho_2 \tag{31}$$

where the indicated sum is over the final particle spins; the quantization axis may be taken along the spin direction of the decaying particle.

a) If the decaying particle has <u>spin zero</u>, then $\Sigma|F|^2$ is independent of angle and from (22) or (29), the total rate is

$$\Gamma = \frac{1}{2M} (\Sigma|F|^2) \frac{p_+}{4\pi M} \tag{32}$$

b) To provide, by way of illustration, <u>a simple model for F</u>, consider the decay of a pseudoscalar K into two pions, with

$$L_{int} = - g \, K \, \pi \, \pi \tag{33}$$

Then F(s) = g (coupling constant), having dimensions of mass. The partial width

$$\Gamma(K \to \pi\pi) = \frac{1}{2} \left(\frac{g^2}{4\pi M^2}\right) p_+ \tag{34}$$

or we may write

$$\Gamma(K \to \pi\pi) = \frac{1}{2} f^2 \, p_+ \tag{35}$$

corresponding to

$$L_{int} = - \sqrt{4\pi} \, M \, f \, K \, \pi \, \pi \tag{36}$$

with dimensionless f.

c) General 2-body decay

A particle at rest of spin j decays into two particles of spin j_1 and j_2 and angular momentum ℓ.

$$\vec{j} = \vec{j}_1 + \vec{j}_2 + \vec{\ell} \tag{37}$$

To get the angular distribution of the decay, combine two of the three final angular momenta using C.-G.coefficients. Then add the third. The number of states contributing is obtained by applying the triangle relations. Often we assume that the lowest allowed ℓ dominates, because of angular momentum barriers. Sometimes it is useful to note that there is no $\vec{\ell}$ in the direction of $\vec{p}_1$ (or $\vec{p}_2$, which is $-\vec{p}_1$) since $\vec{r} \times \vec{p} = 0$.

d) A helicity remark

Consider a high energy electron undergoing a series of interactions, including "backward-in-time" scatterings. For convenience, suppose it to be initially in a state of negative helicity. To the extent that we can neglect its mass, the electron portions of its path in space-time are neutrino-like and the positron portions are anti-neutrino-like. If we define a "generalized helicity" as helicity for electrons and the negative of helicity for positrons, we can prove the generalized helicity is approximately conserved. That is, an interaction which flips the generalized helicity costs a factor m/E in the invariant matrix element.

1. For example, in the decays $\pi^o \to e^+ + e^-$ or $\eta \to e^+ + e^-$, the decay must be a second order electromagnetic interaction, since in either case the electron pair must be in the ^1S_o state, which has charge conjugation number C = +1. (Even if we allow C-violation, we cannot have a first order interaction without violating P. The only vector we can make from π^o is proportional to $\partial\pi^o/\partial x$ which is a pseudovector.)

Likewise, by the arguments given above, the electrons must have equal helicity, and thus opposite generalized helicity. Therefore, the matrix element must contain a factor

$$(m/E) \sim (m_e/m_\pi) \quad .$$

Thus

$$|F|^2 \sim \alpha^4 (m_e/m_\pi)^2 \tag{38}$$

and is therefore very small.

2. The same argument holds, assuming C-conservation, for

$$\eta \to \pi^o + e^+ + e^- \tag{39}$$

providing the π^o is not too fast for the electrons to be relativistic, i.e. the argument holds for almost the entire spectrum.

3. On the other hand, if there is a C-violating part to the conserved vector current (i.e. a C = +1 electric current of η and π^o, in addition to the usual C = -1 current) then, following Bernstein, Feinberg and Lee [2] we write

$$I_\mu = J_\mu + K_\mu \tag{40}$$

where J_μ is the usual C = -1 current and K_μ has C = +1, and furthermore

$$\partial I_\mu / \partial x_\mu = 0 \quad . \tag{41}$$

Applying (41) to

$$\langle \pi^o | I_\mu(x) | \eta^o \rangle = \langle \pi^o | K_\mu(x) | \eta^o \rangle =$$

$$= [f_1(\eta_\mu + \pi_\mu) + f_2(\eta_\mu - \pi_\mu)] \frac{\exp i(\eta_\lambda - \pi_\lambda)x_\lambda}{\sqrt{m_\eta \omega_\pi}} \qquad (42)$$

which is equivalent to multiplying by $q_\mu = \eta_\mu - \pi_\mu$, the latter being momentum 4-vectors, we get

$$f_1(m_\eta^2 - m_\pi^2) = - f_2 q^2 \quad . \qquad (43)$$

So there is only one independent form factor, say f_1. Since we are dealing with a neutral current we consider a form factor like the neutron's, which behaves experimentally like

$$e^{-1/6\, e<r^2>q^2} - 1 \quad . \qquad (44)$$

For small q^2, we take therefore

$$f_1 = - 1/6\, e<r^2>q^2 \qquad (45)$$

where $<r^2>$ is a "size" parameter. This is equivalent to an interaction Lagrangian:

$$L_{int} = e^2/6<r^2>[\eta\partial\pi^o/\partial x_\mu - \pi^o\partial\eta/\partial x_\mu](i\bar{\psi}\gamma_\mu\psi) \qquad (46)$$

and this, in turn, gives for the ratio of the partial rates of decay $\eta^o \to \pi^o e^+ e^-$ and $\eta^o \to 2\gamma$

$$\frac{R(\eta^o \to \pi^o e^+ e^-)}{R(\eta^o \to 2\gamma)} \sim 0.04[<r^2>m_\eta^2]^2 \qquad (47)$$

which is ~ 1, if $<r^2> \sim 1/m_\pi^2$.

We conclude that this ratio is a sensitive test for C-violation. The decay $\eta \to \pi^o e^+ e^-$ has been looked for experimentally but so far not found.

e) The decay $\rho \to 2\pi$

To conclude this section, which uses only elementary methods, we consider the decay via strong interaction

(C,P and I conserving) of the vector particle ρ^o ($J^P = 1^-$) into two pions. Since the two pions must be in a P-state, $\rho^o \rightarrow 2\pi^o$ is forbidden by Bose statistics (and therefore absolutely). It also violates C and I conservation.

Let e_μ be the 4-vector of polarization of the ρ, and k_μ its 4-vector of momentum. Then

$$k_\mu e_\mu = 0 \tag{48}$$

and the invariant matrix element can be written

$$F = g\, e_\mu (p_+ - p_-)_\mu \tag{49}$$

Eq. (48) also tells us that $e_o = 0$ in the rest frame of ρ^o, and since $k_\mu = p_{+\mu} + p_{-\mu}$ we can write (49) as

$$F = 2\, g\, e_\mu\, p_{+\mu} \tag{50}$$

and therefore

$$|F|^2 = 4\, g^2 (\vec{e} \cdot \vec{p}_+)^2 \quad . \tag{51}$$

Suppose ρ^o is polarized along the Z-axis.
Then

$$\vec{e} = \frac{1}{\sqrt{2}}\,(+1, -i, 0) \tag{52}$$

and

$$\vec{e}\cdot\vec{p}_+ = 1/\sqrt{2}\,(p_{+x} - ip_{+y}) = \frac{p_+}{\sqrt{2}} \sin\theta\, e^{-i\phi} \tag{53}$$

so that

$$|F|^2 = 2\, g^2 p_+^2 \sin^2\theta, \qquad \text{for } m = +1 \quad . \tag{54}$$

Similarly, the same result holds for $m = -1$ (i.e., spin

along Z -axis), while for m = 0,

$$|F|^2 = 4\, g^2 p_+^2 \cos^2\theta \qquad (55)$$

For unpolarized ρ^0, we get

$$\frac{1}{3} \sum_{m=0,\pm 1} |F|^2 m = \frac{4}{3}\, g^2 p_+^2 \qquad (56)$$

and finally

$$\Gamma = \frac{g^2 p_+^3}{6\pi M^2} \qquad (57)$$

In this case g is dimensionless.

III. Use of Cartesian Tensors

1. The value of using Cartesian tensors in the phenomenological angular momentum analysis of high energy reactions has been stressed by Zemach [3]. One of the advantages is that it provides a systematic way to construct states of definite orbital angular momentum from a set of particle momenta. Another advantage is that the same techniques can be used to describe the geometry of isospin space or unitary spin space.

In calculating one uses a Cartesian basis, but in interpreting the final result, one often wishes to transform to a spherical basis.

Cartesian basis: $[\vec{n}^1, \vec{n}^2, \vec{n}^3]$ (58)

A unit vector $\vec{p}$ has components (p_1, p_2, p_3), where

$$p_i = \vec{n}^i \cdot \vec{p} \qquad (59)$$

and

$$\vec{p}\cdot\vec{q} = p_1\, q_1 + p_2\, q_2 + p_3\, q_3 \qquad (60)$$

Spherical basis: $|\underline{n}^+, \underline{n}^-, \underline{n}^o|$ (61)

with

$$\underline{n}^{\pm} = \mp \frac{1}{\sqrt{2}} (\vec{n}^1 \pm i\vec{n}^2) \qquad (62)$$

$$\underline{n}^o = \vec{n}^3$$

Spherical components p_M of $\underline{p}$ are given by

$$p_M = (\underline{n}^M)^* \cdot \underline{p} \qquad (63)$$

Thus

$$p_{\pm} = \mp \frac{1}{\sqrt{2}} (p_1 \mp ip_2) = \mp \frac{1}{\sqrt{2}} \sin\theta\ e^{\mp i\phi} \qquad (64)$$

$$p_o = p_3 = \cos\theta \qquad (64a)$$

Note the distinction between basis vectors and components.
The scalar product in the spherical basis is

$$\underline{p}\cdot\underline{q} = p_o q_o - p_+ q_- - p_- q_+ \equiv g^{MM'} p_M q_{M'} \qquad (65)$$

$$g^{MM'} \equiv \begin{array}{c|ccc} & + & o & - \\ \hline + & 0 & 0 & -1 \\ o & 0 & 1 & 0 \\ - & -1 & 0 & 0 \end{array}$$

A completely equivalent alternative formalism, which is more convenient when combining vector wave functions, rather than operators, is to write for the scalar product

$$\underline{p}^* \cdot \underline{q} = p_o q_o + p_+^* q_+ + p_-^* q_- \qquad (66)$$

Both forms, of course, are equal to $p_1q_1 + p_2q_2 + p_3q_3$. Similarly, we use the basis vectors (61) in passing from higher Cartesian to spherical tensors, e.g.:

$$T^{MM'} = (n_i^{M'})^* (n_j^M)^* T_{ij} \tag{67}$$

That is

$$T^{++} = \frac{1}{2}[T_{11} - T_{22} - i(T_{12} + T_{21})]$$

$$T^{+-} = -\frac{1}{2}[T_{11} + T_{22} + i(T_{12} - T_{21})] \quad \text{etc.} \tag{68}$$

We shall be concerned mainly with symmetric traceless tensors, e.g., $T_{ij} - 1/3\ \delta_{ij}\ \mathrm{Tr}\ T$, with $T_{ij} = T_{ji}$. If $T_{ij} = e_i e_j$, $\underline{e}$ being a Cartesian unit vector, one shows easily that the independent components of the spherical tensor are

$$6e_+e_+,\ 6e_+e_o,\ e_oe_o + e_+e_-,\ -6e_-e_o,\ 6e_-e_-,$$

which transform like the spherical harmonics $Y_{\ell m}$, for $\ell = 2$.

2. A particle of spin j is represented in its rest frame by a symmetrical, traceless tensor (a so-called "pure" tensor) of rank j:

$$T^j_{m_1 m_2 \ldots m_j} \qquad m_i = 1,2,3. \tag{69}$$

This can be shown to have 2j + 1 independent components. The symmetry is expressed by

$$T^j_{\ldots m_i \ldots m_k \ldots m_j} = T^j_{\ldots m_k \ldots m_i \ldots m_j}\ , \tag{70}$$

each pair i, k.

The tracelessness condition for a symmetric tensor is simply

$$\Sigma_m T^j_{mm\ldots m_j} = 0 \tag{71}$$

Such a tensor is equivalent to an angular momentum wave function Y^j_m. Suppose the tensor indices $m_1 \ldots m_j$ are expressed in the spherical representation. In such a case one can write

$$T^j_{m_1\ldots m_j} = T^j_{M_o M_+ M_-} \quad ,$$

where M_+ is the number of + indices, etc. Then only the numbers M_o, M_+, M_- are important, since the magnetic quantum number

$$m = M_+ - M_- \quad . \tag{72}$$

Example: One can make symmetric tensors by using identical vectors in tensor products (i.e. combining unit spins). The first few symmetric traceless (irreducible) tensors are:

$$\begin{array}{ll}
J = 0 & 1 \\
1 & \underline{a} = (a_+, a_o, a_-) \\
2 & a_i a_j - \frac{1}{3} \underline{a}^2 \delta_{ij} \\
3 & a_i a_j a_k - \frac{1}{5}(\delta_{ij} a_k + \delta_{ik} a_j + \delta_{jk} a_i)\, \underline{a}^2 \\
4 & a_i a_j a_k a_l - \frac{1}{7} \Sigma_P\, \delta_{ij} \underline{a}^2 a_k a_l - \frac{1}{35} \cdot \\
 & \cdot\, (\delta_{ij}\delta_{kl} + \delta_{ik}\delta_{jl} + \delta_{il}\delta_{jk})(\underline{a}^2)^2 \quad ,
\end{array} \tag{73}$$

where Σ_P means sum over permutations.

3. A useful formula (Zemach [3])

If a tensor is built from a single vector, write it

$T^j(\underline{p})$. The scalar product of two such tensors of equal rank is ($\underline{n}$ and $\underline{p}$ being unit vectors)

$$T^j(\underline{n}) : T^j(\underline{p}) = C_j P_j(\underline{n}.\underline{p}) \tag{74}$$

where $P_j(x)$ is the Legendre polynomial. From $P_j(1) = 1$, one gets

$$C_j = \frac{j!}{(2j - 1)!!} \tag{74a}$$

$$[(2j - 1)!! = (2j - 1)(2j - 3)\ldots]$$

We have,e.g.,

$$P_3(x) = \frac{5}{2}(\underline{n}\ \underline{n}\ \underline{n} - \frac{3}{5}\ \delta\underline{n}) : (\underline{p}\ \underline{p}\ \underline{p}) = \frac{1}{2}(5x^3 - 3x),$$

$$x = \underline{n} \cdot \underline{p} \quad .$$

We have used here the fact that only one tensor in the scalar product need be pure (i.e. traceless symmetric), since the procedure for making a tensor pure can be expressed as a projection operation.

An Application

Consider a particle of spin j decaying into two spinless particles. The parities are not important. The only direction associated with the initial state is the direction $\underline{n}$ of the spin vector, and the only direction in the final state is $\underline{p}$, the decay momentum. The interaction operator can depend only on the one relative direction.

Thus, using (31),

$$d\Gamma = \frac{p}{8\pi M^2}|F|^2 \frac{d\Omega}{4\pi} \tag{75}$$

and

$$F = g\langle\phi_m^j(\underline{n})|A^j(\underline{p})|\phi^o\rangle \tag{76}$$

and by (74)

$$F = g\ p^j C_j^{1/2} P_j(\cos\theta). \tag{77}$$

The reason for $C_j^{1/2}$ is that $|\phi_m^j\rangle$ is normalized to 1, while A^j is normalized to C_j (see (74)), so

$$\phi^j : A^j = C_j^{1/2} P_j(\cos\theta) \quad .$$

Thus

$$d\Gamma = \frac{g^2 p^{2j+1}}{8\pi M^2}\ C_j P_j^2(\cos\theta)\ \frac{d\Omega}{4\pi} \quad . \tag{78}$$

Since $\Gamma = \int d\Gamma$, and

$$\int P_j^2(\cos\theta)\ \frac{d\Omega}{4\pi} \equiv (2j + 1)^{-1} \ , \tag{79}$$

we get

$$\Gamma = \frac{g^2 p^{2j+1}}{8\pi M^2}\ \frac{C_j}{(2j + 1)} \quad . \tag{80}$$

Alternatively, we can average $|F|^2$ over spins, giving

$$\Gamma = \frac{p}{8\pi M^2}\ g^2 A^j(p) : A^j(p) \int\frac{d\Omega}{4\pi}(2j + 1)^{-1} \ , \tag{81}$$

which leads once again to (80).

In any case, we get

$$\Gamma = g^2\ \frac{p^{2j+1}}{8\pi M^2}\ \frac{j!}{(2j + 1)!!} \tag{82}$$

or

$$\Gamma = \frac{1}{2\pi}\ f^2 p^{2j+1}\ \frac{j!}{(2j+1)!!} \ , \qquad f^2 = \frac{g^2}{4\pi M^2} \quad . \tag{83}$$

For j = 0, f^2 is dimensionless.
For j = 1, g^2 is dimensionless.

In the general case f^2 has dimensions $(\text{mass})^{-2j}$ and it would be more convenient to write

$$f^2 = G^2 \mu^{-2j} \quad ,$$

where G^2 is dimensionless.

If we are dealing with isospin or SU(3), etc., multiplets we write finally

$$\Gamma = \frac{G^2}{2\pi} \mu (\frac{p}{\mu})^{2j+1} \frac{j!}{(2j+1)!!} |a_{\gamma_1 \gamma_2 \gamma_3}|^2 \quad , \tag{84}$$

(a formula also given by Behrends).

Here $a_{\gamma_1\gamma_2\gamma_3}$ is a suitable defined C. G. coefficient for

$$\gamma_3 \rightarrow \gamma_1 + \gamma_2$$

and μ is a charakteristic mass.

IV. Isospin Problems and SU(3), etc., Problems

1. Isospin

Consider the problem of combining two isospins I_1 and I_2 to make I. One can use the tensor methods described above as for angular momentum. However in this case the result is given directly in the desired (i.e. spherical) form by Clebsch-Gordan coefficients. For lower spins the methods are, perhaps, equally convenient. For higher spins C.G. coefficients are preferable. The tensor method has, however, one important advantage.

Example: From two spins one $\underline{a}_1$, $\underline{a}_2$ we can make

$$I = 0 \quad \underline{a}_1 \cdot \underline{a}_2 \tag{85}$$

$$1 \quad \underline{a}_1 \times \underline{a}_2 \tag{86}$$

$$2 \quad \frac{1}{2}[\underline{a}_1\underline{a}_2 + \underline{a}_2\underline{a}_1] - \frac{1}{3}\,\underline{a}_1\underline{a}_2\, I \ . \tag{87}$$

The advantage is that for identical particles the symmetry is obvious. The same is true for combining several isospins.

Example: Three pions $\underline{a}_1$, $\underline{a}_2$, $\underline{a}_3$ as occur in $\eta \rightarrow 3\pi$, $K \rightarrow 3\pi$, etc.

$$I = 0 : \underline{a}_1 \times \underline{a}_2 \cdot \underline{a}_3 \quad \text{(antisymmetric)} \tag{88}$$

Thus there is no $3\pi^o$ decay (triple product is a determinant in the spherical components). To satisfy Bose statistics, this must be multiplied by an antisymmetric function of the energy-momentum variables. For η-decay this state violates C; for K_2-decay it violates CP.

I = 1: Triple vector product, but this is equivalent to

$$A_1\, \underline{a}_1\, (\underline{a}_2 \cdot \underline{a}_3) + A_2\, \underline{a}_2\, (\underline{a}_3 \cdot \underline{a}_1) + A_3\, \underline{a}_3\, (\underline{a}_1 \cdot \underline{a}_2) \tag{89}$$

where the A's depend on energy-momentum variables.

To satisfy Bose statistics A_1 is symmetric in pions 2, 3 and analogously for A_2 and A_3.

If (89) is symmetric in pions 2, 3 then also $A_2 = A_3$. For $3\pi^o$ we can conclude the matrix element is $A_1 + A_2 + A_3$. For pions 1,2,3 being π^+, π^-, π^o the matrix element is $-A_3$. Etc.

A similar analysis can be made for I = 2 (see Zemach). For I = 1 and I = 2 the symmetry in isospin is not symmetric or antisymmetric, but mixed. For I = 3 we get the third rank symmetric traceless tensor (the highest isospin that can be formed is always symmetric).

ΔI Rules and Spurions

Consider a decay like $K \to 2\pi$ in which $I = 1/2$ goes to two $I = 1$ particles, and which therefore violates I-spin conservation. We can treat this as an I-spin conserving process by introducing a spurion S which carries only the I-spin violation (and not charge, mass,energy, etc.). The only weak interaction spurion which appears as absolutely necessary to date is $\Delta I = 1/2$. For $K \to 2\pi$ we have

$$I + \Delta I = \underline{1/2} + \underline{1/2} = \underline{0} \text{ or } \underline{1} \quad . \tag{90}$$

The electromagnetic current has $I = 1$, so a photon interaction has $\Delta I = 1$, and therefore if a virtual photon is involved in a process (two interactions) it can give $\Delta I = 0$, 1, or 2. Thus, by use of spurions, we can always use a charge independent formalism. Thus, for example, when we write the Lagrangian

$$L = g \bar{N} N \pi \tag{91}$$

this implies a set of terms involving different members of the I-spin multiplets involved. More explicitly we mean

$$i G(\bar{N} \underline{\tau} . N) \underline{\phi} = i G[\sqrt{2} \bar{p} n \pi_+ + \sqrt{2} \bar{n} p \pi_- + + (\bar{p}p - \bar{n}n)\pi_o] \quad . \tag{92}$$

Similarly, for the Lagrangian (with implied spurion)

$$L = -g K \pi \pi \tag{93}$$

In this case, since the K has spin zero only the symmetric S-state is possible for the two pions, and thus

only I = 0. In the tensor notation the form of the matrix element is $\underline{a}_1 \cdot \underline{a}_2$.
Since

$$\underline{a}_1 \cdot \underline{a}_2 = a_o a_o - a_+ a_- - a_- a_+ \tag{94}$$

When we take into account the identity of the π_o's, we conclude

$$\frac{\Gamma(\pi^+\pi^+)}{\Gamma(\pi^o\pi^o)} = \frac{2}{1} \tag{95}$$

It is important to make a careful interpretation of (94) to obtain this result. Let us spell it out in detail with Feynman diagrams:

For $\pi^+ + \pi^-$

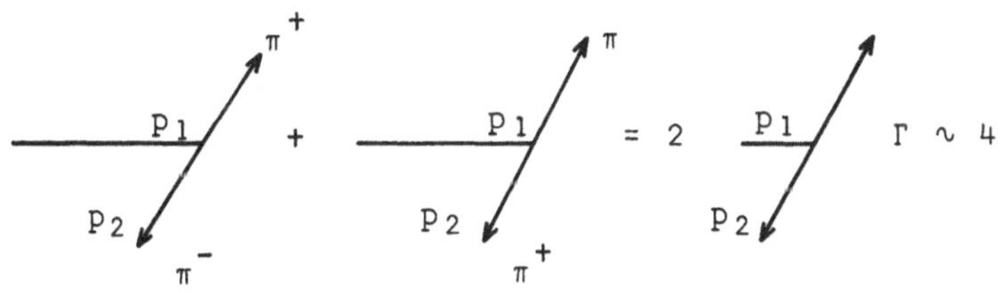

For $\pi^o + \pi^o$

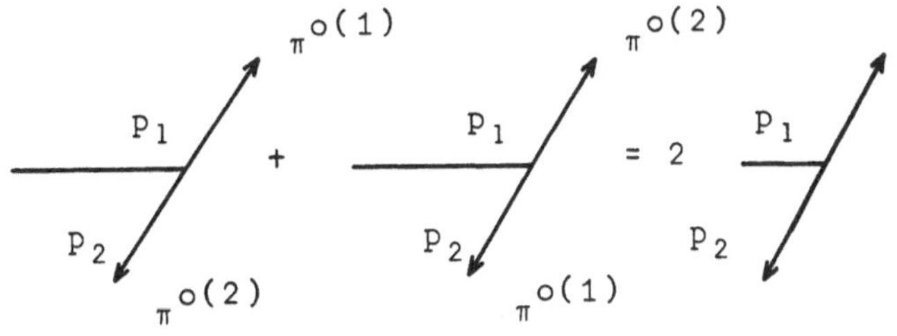

but only half the phase space is available, so $\Gamma \sim 2$. Alternatively, use only one diagram for $\pi^o + \pi^o$ but symmetrize the wave function

$$\frac{\pi_o(1)\,\pi_o(2) + \pi_o(2)\,\pi_o(1)}{\sqrt{2}}$$

In this case use the whole phase space.

2. SU(3) Problems: How to use the tables (de Swart [4])

This is a very condensed summary of the important points. For details see de Swart.

Consider a three-dimensional complex vector space with vectors x^i, and let $(x^i)^* = x_i$. SU(3) is the group of all unitary unimodular (i.e. unit determinant) transformations in this space.

Thus,

$$\bar{x}^i = a_{ij}\, x^j \tag{96}$$

and

$$\bar{x}_i = a^{-1}_{ji} x_j = a^*_{ij} x_j \text{ , since } a^+ = a^{-1} . \tag{97}$$

A tensor

$$A^{\alpha\beta\ldots\delta}_{ij\ldots l} \tag{98}$$

with p upper and q lower indices transforms like a product of vectors

$$x^\alpha y^\beta \ldots z^\delta u_i v_j \ldots w_l \tag{99}$$

An irreducible representation (p,q) is represented by a tensor having p upper and q lower indices, totally symmetric in the upper indices and totally symmetric in the lower indices and traceless, i.e.

$$\sum_\alpha A^{\alpha\beta\ldots\delta}_{\alpha j\ldots l} = 0 \tag{100}$$

Such an irreducible representation has dimension (i.e. number of independent components)

$$N = (1 + p)(1 + q) \left[1 + \frac{(p+q)}{2}\right] \tag{101}$$

E.g. the representation (1,1) has N = 8.
Different 8's may be distinguished by quantum numbers other than p,q.

A member of an SU(3) representation is labeled by the quantum numbers

$$[p, q, I^2, I_z, Y] \tag{102}$$

which are all related to generators of the group.

If we form the product of two IR's, the product representation is generally reducible; however, it can be uniquely decomposed into the sum of IR's. A product state is then specified by the ten quantum numbers

$$[p, q, I^2, I_z, Y]_1 \quad \text{and} \quad [p, q, I^2, I_z, Y]_2 \ .$$

From the sum of the generators (total unitary spin)

$$F_i = F_i^{(1)} + F_i^{(2)}, \qquad i = 1...8 \tag{103}$$

we can form nine quantum numbers:

$$p_1, q_1, p_2, q_2, p, q, I^2, I_z, Y \quad . \tag{104}$$

This implies a degeneracy which can be removed only by an "external" quantum number Γ, which is not in the group.

Let

$$\mu_i = (p,q)_i \tag{105}$$

$$\gamma_i = (I^2, I_z, Y)_i \tag{106}$$

Then a state of the irreducible product representation γ is given by

$$[\mu_\gamma , \gamma] = \psi \begin{pmatrix} \mu_1 & \mu_2 & \mu_\gamma \\ & & \gamma \end{pmatrix} . \qquad (107)$$

The reducible and irreducible product representations are connected by C. G. coefficients:

$$\psi \begin{pmatrix} \mu_1 & \mu_2 & \mu_\gamma \\ & & \gamma \end{pmatrix} = \sum_{\gamma_1, \gamma_2} \begin{pmatrix} \mu_1 & \mu_2 & \mu_\gamma \\ \gamma_1 & \gamma_2 & \gamma \end{pmatrix} \phi_{\gamma_1}^{(\mu_1)} \phi_{\gamma_2}^{(\mu_2)} . \qquad (108)$$

By using the usual C.G. coefficients (i.e. those of SU(2)) to combine the products of states into eigenstates of I^2, we can write

$$\begin{pmatrix} \mu_1 & \mu_2 & \mu_\gamma \\ \gamma_1 & \gamma_2 & \gamma \end{pmatrix} = C^{I_1 \; I_2 \; I}_{I_{1_z} \; I_{2_z} \; I_z} \left(\begin{matrix} \mu_1 & \mu_2 \\ I_1 Y_1 & I_2 Y_2 \end{matrix} \middle| \begin{matrix} \mu_\gamma \\ IY \end{matrix} \right) . \qquad (109)$$

The last factor is called an "isoscalar factor", and these are tabulated. In the phases of the C's the Condon-Shortley convention is used. The phases of the product states are taken real. Highest I_1, then highest I_2 where necessary, is given a positive C.G.coefficient.

To show how the tables are used, we quote a portion of the de Swart's table and then give a physical example of its use.

Isoscalar Factors for

$\underline{8} \times \underline{8} \rightarrow \underline{8}$

	I	Y	I_1	Y_1	I_2	Y_2	$\underline{8}_1$ $\square\square$ (D)	$\underline{8}_2$ $\begin{smallmatrix}\square\\\square\end{smallmatrix}$ (F)
(η)	0	0	1/2	1	1/2	-1	$\sqrt{10}/10$	$\sqrt{2}/2$
			1/2	-1	1/2	1	$-\sqrt{10}/10$	$\sqrt{2}/2$
			1	0	1	0	$-\sqrt{15}/5$	0
			0	0	0	0	$-\sqrt{5}/5$	0
(π)	1	0	1/2	1	1/2	-1	$-\sqrt{30}/10$	$\sqrt{6}/6$
			1/2	-1	1/2	1	$-\sqrt{30}/10$	$-\sqrt{6}/6$
			1	0	1	0	0	$\sqrt{6}/3$
			1	0	0	0	$\sqrt{5}/5$	0
			0	0	1	0	$\sqrt{5}/5$	0
(K)	$\frac{1}{2}$	1	1/2	1	1	0	$3\sqrt{5}/10$	1/2
			1	0	1/2	1	$-3\sqrt{5}/10$	1/2
			1/2	1	0	0	$-\sqrt{5}/10$	1/2
			0	0	1/2	1	$-\sqrt{5}/10$	-1/2
($\bar{K}$)	$\frac{1}{2}$	-1	1/2	-1	1	0	$-3\sqrt{5}/10$	1/2
			1	0	1/2	-1	$3\sqrt{5}/10$	1/2
			1/2	-1	0	0	$-\sqrt{5}/10$	-1/2
			0	0	1/2	-1	$-\sqrt{5}/10$	1/2

As an application we consider the decay into two members of the pseudoscalar octet of a proposed scalar (0^+) octet [5]:

	I	Y
$\varepsilon^0(640)$	0	0
ξ (950)	1	0

	I	Y
K (725)	1/2	1
$\bar{K}$ (725)	1/2	-1

(The ε^o is conjectured to be a mixture of two I = 0 physical particles ε and σ.) Since the meson octet is self-conjugate, i.e. the octet is its own anti-octet, we use the isoscalar factors for $\underline{8} \times \underline{8} \to \underline{8}_1$, corresponding to symmetric or D-coupling.

Constructing the states (I, Y):

$$\varepsilon^o(0,0) = -\frac{\sqrt{15}}{5}\,\pi(1)\,\pi(2) + \bar{K}K \text{ and } \eta\eta \text{ terms}$$

of which only the $\pi\pi$ channel is open. We consider further only the open channels.

$$\xi(1\ \ ,\ 0) = \frac{\sqrt{5}}{5}\left[\eta(1)\,\pi(2) + \pi(1)\,\eta(2)\right]$$

$$K(1/2,\ 1) = \frac{3\sqrt{5}}{10}\left[K(1)\,\pi(2) - \pi(1)\,K(2)\right]$$

$$\bar{K}(1/2,-1) = -\frac{3\sqrt{5}}{10}\left[\bar{K}(1)\,\pi(2) - \pi(1)\,\bar{K}(2)\right]$$

To compare the partial decay rates into two 0^- mesons of, say K^+ and ξ^o we need only to take squares of the isoscalar factors:

$$\Gamma_{K^+}/\Gamma_{\xi^o} = \frac{9}{5}/\frac{4}{5} \times \text{kinematical ratio.} \qquad (110)$$

But since the identity of particles is not taken into account in C. G. coefficients or in isoscalar factors we must be more careful for ε^odecay. Namely, we can count only one-half of the $\pi^o\pi^o$ contribution.

That is, we must interpret $\pi(1)\,\pi(2)$ as

$$\frac{1}{\sqrt{3}}\left[\pi^+(1)\,\pi^-(2) + \pi^-(1)\,\pi^+(2) - \frac{\pi^o(1)\pi^o(2)+\pi^o(2)\pi^o(1)}{\sqrt{2}}\right]$$

$$= \frac{1}{\sqrt{3}} \; [2\pi^+\pi^- - \sqrt{2}\pi^o\pi^o] \quad , \tag{111}$$

and the appropriate squared coefficient for Γ_{ε^o} is

$$\left(\frac{\sqrt{15}}{5}\right)^2 \times \left(\frac{4}{3} + \frac{2}{3}\right) = \frac{6}{5} \quad . \tag{112}$$

If $\pi(1)$ and $\pi(2)$ were never identical (i.e. if $\pi^o(1)$ and $\pi^o(2)$ were not identical), we would have obtained 12/5. For a discussion of symmetry breaking effects, appropriate at this point, see the review article of de Swart.

V. Meson Decays Involving Photons

1. General remarks

Questions of gauge invariance can be avoided by using instead of A_μ the fields

$$F_{\mu\nu} = k_\nu \, A_\mu - k_\mu \, A_\nu \tag{113}$$

Introducing a transverse polarization three-vector $\underline{e}$, this breaks into the usual

$$\underline{E} = k_o \, \underline{e} \quad , \quad \underline{E} \cdot \underline{k} = 0 \; , \quad \text{parity odd} \tag{114a}$$

and

$$\underline{H} = \underline{k} \times \underline{e} \; , \quad \underline{H} \cdot \underline{k} = 0, \quad \text{parity even} \tag{114b}$$

From these vectors one can build symmetric traceless tensors [3] as in Section III:
electric multipoles (2j-poles) of parity $(-1)^j$

$$k_o T^j(\underline{k}\ \underline{k}\ldots\underline{k}.\underline{e}) \tag{115}$$

and magnetic multipoles (2j-poles) of parity $(-1)^{j+1}$

$$T^j(\underline{k}\ \underline{k}\ldots\underline{k}\ \underline{k}\times\underline{e}) \quad . \tag{115b}$$

Consider now a meson decay

$$A \to B + \gamma \tag{116}$$

where A is a massive meson at rest. If we make a helicity measurement of the photon we must get +1 or -1 and conclude, since the orbital angular momentum component along $\underline{k}$ must vanish:

a) At least one of the particles A or B has spin (i.e. zero to zero transitions are forbidden).

b) If A is spinless, B and γ have the same helicity. If B is also a photon, the two photons are both right or both left circularly polarized (Yang). In addition their phases are correlated for 0^+ or 0^- decay.

We can also prove:

c) If A has spin 1, B cannot be a photon, i.e. spin $1 \to 2\gamma$ is forbidden.

Proof: In the rest system of the decaying particle consider orthogonal unit vectors $\hat{\underline{k}}$, $\underline{e}^{(1)}$, $\underline{e}^{(2)}$.

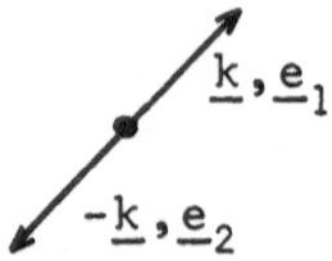

The photon polarizations are

$$\underline{e}_i = a_{i1}\,\underline{e}^{(1)} + a_{i2}\,\underline{e}^{(2)}, \qquad i = 1,2 \tag{117}$$

and thus we need consider only $\underline{e}^{(1)}$, $\underline{e}^{(2)}$ (four cases). The state of each photon is represented by a tensor of the form

$$T^m = T^m(\underline{\hat{k}}\ldots\underline{\hat{k}}\ e^{(j)}),\quad m = \ell\ ,\ \ell \pm 1 \tag{118}$$

since $\underline{\hat{k}} \times \underline{e}^{(1)} = \underline{e}^{(2)}$ and $-\underline{k} \times \underline{e}^{(2)} = \underline{e}^{(1)}$. (119)

The two tensors must be contracted to a vector (for J = 1) which is symmetric in the photon exchange, i.e., either

$$\underline{k}(\underline{e}^{(1)}\cdot\ \underline{e}^{(2)}) \qquad \text{or } \underline{k}\ . \tag{120}$$

But $\underline{e}^{(1)}\cdot\ \underline{e}^{(2)} = 0$, while $\underline{k} = 1/2\ (\underline{k}_1 - \underline{k}_2)$

which is antisymmetric in exchange. (For a different proof see Reference 1).

d) By the same reasoning we can also see immediately that for

$$0^+ \text{ decay}\ ,\qquad M \sim \underline{e}_1 \cdot \underline{e}_2 \tag{121}$$

$$0^- \text{ decay}\ ,\qquad M \sim \underline{e}_1 \times \underline{e}_2 \cdot \underline{k} \tag{122}$$

e) For two photon decay, when we put in the kinematic factors we obtain from (22) and (3):

$$d\Gamma = \frac{k}{2\pi M^2}\ |F|^2 \frac{d\Omega}{4\pi} \tag{123}$$

and as k = M/2

$$d\Gamma = \frac{1}{4\pi M}\ |F|^2\ \frac{d\Omega}{4\pi}\ . \tag{124}$$

We know F must be proportional to αk, but to define a coupling strength we must have a model. E.g.,

$$L = \frac{\alpha g}{M}\ \varepsilon^{\mu\nu\lambda\sigma}\ p_\mu k_\nu e_\lambda e_\sigma \tag{125}$$

is a possible Lagrangian.

2. The Gell-Mann - Sharp - Wagner Model [6]

This model assumes the existence of a transition

vector meson $\longleftrightarrow$ photon

as it occurs in the model of the electromagnetic form factor. Indeed the success of this model for the form factor led to the experimental search for the ρ-meson.

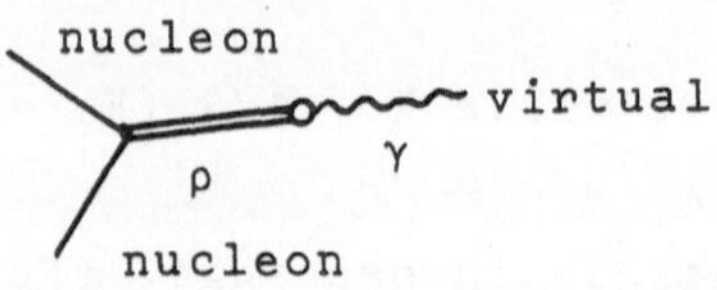

The neutral vector mesons ρ^o, ω, and ϕ are all possible contributors to this process. Once the V-γ couplings are determined, the remaining couplings in the GSW model are those of the strong interactions.

If the vector mesons dominate the charge form factors (isovector and isoscalar) of the nucleon, then in the limit of low energy photons we have

$$f_{V\gamma} = \frac{em_V^2}{2f_V} \tag{126}$$

where f_V is the Yukawa coupling of the vector meson to the nucleon, in order that the charge form factors for low momentum transfer be e/2.

In pure SU(3) theory a U-spin subgroup can be defined as an analogue of I-spin. The U-spin multiplets consist of particles of the same electric charge, and the emission of a photon does not, of course, change the charge. Thus while the photon creation and annihilation operators have mixed I = 0 and I = 1 character, they are U-spin scalars.

Therefore the coupling of the photon to the unmixed and unbroken vector octet (i.e. to ω_8 and ρ^o) is to the U-spin scalar combination which is $\omega_8 + \sqrt{3}\,\rho^o$.

If we write this coupling as

$$L = \frac{f}{\sqrt{3}} V_\mu A_\mu \qquad (127)$$

then

$$f_{\omega_8\gamma} = \frac{f_{\rho\gamma}}{\sqrt{3}} \qquad (128)$$

Now it is known, of course, that the SU(3) symmetry of the vector octet is broken by mass splittings, and also it is thought to be mixed with the SU(3) singlet ω_1. It is not entirely clear what to do about this. However, a possible, and not unreasonable, procedure has been proposed by Dashen and Sharp [7].

Start with the mixing formulation of Sakurai and write

$$|\omega\rangle = a|\omega_8\rangle + b|\omega_1\rangle$$

$$|\phi\rangle = -b|\omega_8\rangle + a|\omega_1\rangle \qquad (129)$$

with a,b real and $a^2 + b^2 = 1$;
and imagine $|\omega\rangle$ and $|\phi\rangle$ to be the physical particles <u>with the physical masses</u>. Although $|\omega_1\rangle$ is itself a U-spin scalar, we assume it is not coupled to the photon [8]. We get then, from (126) and (129):

$$f_{\omega\gamma} = \frac{a\, m_\omega^2}{\sqrt{3}\, m_\rho^2} f \qquad (130)$$

$$f_{\phi\gamma} = - \frac{b\, m_\phi^2}{\sqrt{3}\, m_\rho^2} f \quad , \qquad (131)$$

with

$$f_{\rho\gamma} = f . \qquad (132)$$

Thus the photon coupling to the vector nonet is

$$L_{int} = -\frac{f}{\sqrt{3}m_\rho^2}\{\sqrt{3}\, m_\rho^2 \rho_\mu + a\, m_\omega^2\, \omega_\mu - b\, m_\phi^2\, \phi_\mu\}A_\mu . \qquad (133)$$

Here $a = \cos\theta$, $b = \sin\theta$, and $\theta \simeq 50^\circ$.
Then, using (126) and assuming $f_\rho = f_{\rho\pi\pi}$ we get from a ρ-width $\Gamma_\rho \simeq 106$ MeV (using(57)),

$$\frac{f_\rho^2}{4\pi} \simeq 2.2 \quad . \qquad (134)$$

And using (126), we get

$$4\pi f^2 = \frac{e^2 m_\rho^4}{8.8} \quad . \qquad (135)$$

This is one way to determine the vector meson-photon couplings.

As an example consider π^0 decay [9] :

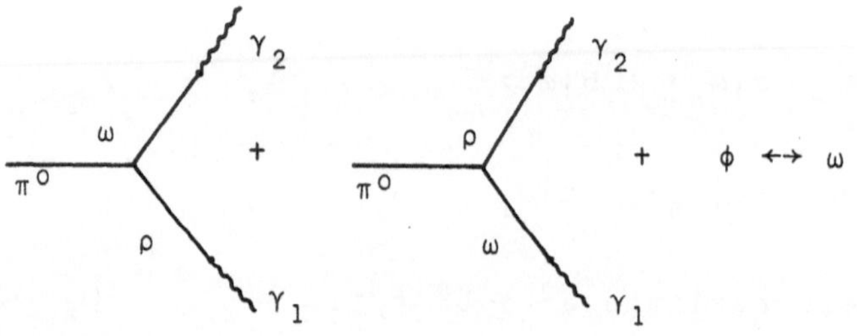

The matrix element is

$$M = G\, \varepsilon^{\alpha\beta\gamma\delta} \varepsilon_\alpha^{(2)}\; k_\beta^{(2)}\; P_\gamma\; e_\delta\; e_\mu\; \varepsilon_\mu^{(1)} \qquad (136)$$

where the ε's are photon polarization vectors, and the e's are vector meson polarizations; P_γ is the four momentum of the decaying pion, and G is given by

$$G = \frac{2f_{\pi\rho\omega}\; f_{\rho\gamma}\; f_{\omega\gamma}}{m_\omega^2\; m_\rho^2} + \frac{2f_{\pi\rho\phi}\; f_{\rho\gamma}\; f_{\phi\gamma}}{m_\phi^2\; m_\rho^2} \quad . \qquad (137)$$

If we denote the $P_8 V_8 V_8$ coupling by g and the $P_8 V_8\, \omega_1$ coupling by g', SU(3) gives

$$f_{\pi\rho\omega} = \frac{a}{\sqrt{5}} g + b g' \qquad (138)$$

$$f_{\pi\rho\phi} = - \frac{b}{\sqrt{5}} g + a g' \qquad (139)$$

Putting these in, we get (with (130), (131), (132))

$$G = \frac{2f^2 g}{m_\rho^4 \sqrt{15}} \qquad (140)$$

From the known π^o lifetime we thus determine g, since this result is independent of g' (and also of a and b). The same is true for $\eta \to 2\gamma$ and $\eta \to \pi^+\pi^-\gamma$ for which, therefore, we can determine (for $\Gamma(\pi^o \to 2\gamma) = 6.2$ eV)

$$\Gamma(\eta \to 2\gamma) = 152 \text{ eV}$$

$$\Gamma(\eta \to \pi\pi\gamma) = 36 \text{ eV} \quad .$$

The ratio 4.2 can be compared with the experimental ratio 4.9 ± 0.8 .

Note: Substantially the same set of lectures has been given previously in 1966 at the Second Finish Summer School in Physics, arranged by the Research Institute for Theoretical Physics at the University of Helsinki.

References

1. K. Nishijima, "Fundamental Particles" (W. A. Benjamin, Inc., 1963).
2. J. Bernstein, G. Feinberg and D. T. Lee, Phys. Rev. 139, B1650 (1965).
3. C. Zemach, Phys. Rev. 140, B 97 and B 109 (1965); also Phys. Rev. 133, B 1201 (1964).
4. J. J. de Swart, Rev. Mod. Phys. 35, 916 (1963).
5. L. M. Brown, Phys. Rev. Lett. 14, 836 (1965).
6. M. Gell-Mann, D. Sharp and W. Wagner, Phys. Rev. Lett.

8, 261 (1962).
7. R. F. Dashen and D. H. Sharp, Phys. Rev. 133, B1585 (1964).
8. J. J. Sakurai, Ann. Phys. (N.Y.) 11, 1 (1960).
9. H. Faier, Nuovo Cimento, 41, 127 (1965).

DISPERSION SUM RULES IN ELEMENTARY PARTICLE PHYSICS†

By

V. de ALFARO

Istituto di Fisica Teorica
dell'Università, Torino

1. Introduction

In these notes I shall try to give a short account of the present status of the strong interaction sum rules. Since the field is in rapid expansion, I apologize for the general and applicative works I am not able to review. The last section is devoted to some observations about the analogous sum rules arising in current algebra.

Of course that section is not meant to be a review of current algebra, where many other important problems and beautiful consequences arise (as for instance the low energy theorems), but only to be a short comparison with the strong interaction sum rules, stressing similarities and differences.

The unpublished results included in the lectures have been obtained in collaboration with my friends S. Fubini, G. Furlan and C. Rossetti, whom I wish to thank for the help in preparing these notes.

---

† Lecture given at the VI. Internationalen Universitätswochen f.Kernphysik,Schladming,26 February-11 March 1967.

2. Asymptotic Behaviour of Scattering Amplitudes for Large Energy and Fixed Momentum Transfer

The possibility of establishing sum rules for amplitudes describing scattering of strongly interacting particles depends upon the knowledge of their high energy behaviour plus the usual analyticity requirements [1].

Let us consider an amplitude $f(\nu)$ satisfying the usual analyticity requirements with the asymptotic behaviour for large ν:

$$|f(\nu)| < C\,\nu^{b}\ ,\ b < 0. \tag{2.1}$$

Then the non subtracted dispersion relation holds

$$\mathrm{Re}\, f(\nu) = \frac{1}{\pi} P\int \frac{\mathrm{Im}\, f(\nu')}{\nu'-\nu}\, d\nu' \tag{2.2}$$

Suppose now that, in (2.1), $b < -1$. We say, in this case, that the amplitude $f(\nu)$ is superconvergent and we obtain the condition

$$\int \mathrm{Im}\, f(\nu) d\nu = 0 \tag{2.3}$$

from

$$\lim_{\nu\to\infty} \nu |f(\nu)| = 0 \ .$$

Adopting a Regge pole point of view, we would need a channel dominated by a trajectory with $\alpha < -1$ to get superconvergent amplitudes in a scattering of scalar particles. The situation changes entirely when we are in presence of particles endowed with spin different from zero. In that case we shall see that further convergence factors are derived as a consequence of the presence of spins, which makes it possible to have superconvergent amplitudes without requiring such a small value of α.

Let us consider in detail the $\rho + \pi \to \rho + \pi$ scattering. We expand the invariant matrix T as follows (kinematics is defined in fig. 2.1):

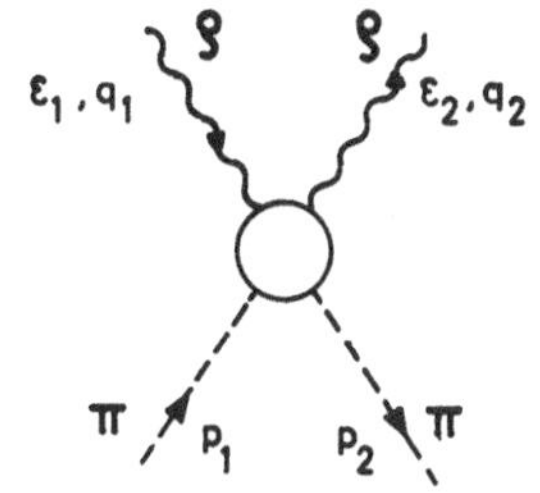

Fig. 2.1

$$u = (p_1 - q_2)^2$$
$$s = (p_1 + q_1)^2$$
$$\Delta = p_2 - p_1 = q_1 - q_2$$
$$t = \Delta^2$$
$$P = p_1 + p_2$$
$$Q = q_1 + q_2$$
$$\nu = \frac{1}{2}(P\,Q) = \frac{1}{2}(s - u)$$

$$T = (\varepsilon_1 P)(\varepsilon_2 P)A + \left[(\varepsilon_1 P)(\varepsilon_2 Q) + (\varepsilon_2 P)(\varepsilon_1 Q)\right]B + $$
$$+ (\varepsilon_1 Q)(\varepsilon_2 Q)C_1 + (\varepsilon_1 \varepsilon_2)C_2 \quad . \qquad (2.4)$$

The invariants A, B, C_1, C_2 are functions of ν and t. Let us apply a Regge pole model to determine the asymptotic behaviour of A, B, $C_{1,2}$ for large ν at fixed t. We consider an exchange of a particle of spin J and parity $(-1)^J$ in the t channel according to the graph of fig. 2.2 :

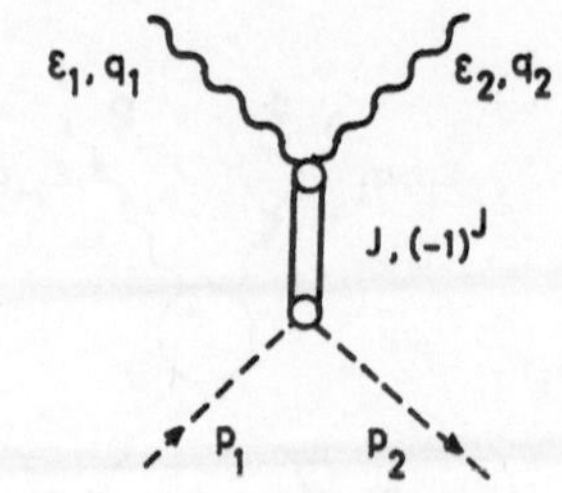

Fig. 2.2

To simplify kinematics we shall take all masses equal. This is of no importance in establishing the asymptotic behaviours.

Describing the intermediate particle through a field $\Phi_{\mu_1 \ldots \mu_J}$ symmetrical in the indices, traceless and obeying transversality conditions, we may write the couplings

$$\begin{aligned}\rho\rho J) \quad & [f_1^{(J)}(\varepsilon_1 \varepsilon_2) + f_2^{(J)}(\varepsilon_1 Q)(\varepsilon_2 Q)]\ \Phi_{\mu_1 \ldots \mu_J} Q_{\mu_1} \ldots Q_{\mu_J} + \\ & + f_3^{(J)}[\varepsilon_{1\mu_1}(\varepsilon_2 Q) + \varepsilon_{2\mu_1}(\varepsilon_1 Q)]\Phi_{\mu_1 \ldots \mu_J} Q_{\mu_2} \ldots Q_{\mu_J} + \\ & + f_4^{(J)}\ \varepsilon_{1\mu_1} \varepsilon_{2\mu_2} \Phi_{\mu_1 \ldots \mu_J} Q_{\mu_3} \ldots Q_{\mu_J} \quad , \end{aligned} \tag{2.5}$$

$$\pi\pi J) \quad g^{(J)}\ \Phi_{\mu_1 \ldots \mu_J} P_{\mu_1} \ldots P_{\mu_J} \tag{2.6}$$

We may perform the sum over polarizations of the Φ particle, getting

$$\begin{aligned} F_J &= \sum_{\text{pol.}\alpha} \Phi^{(\alpha)}_{\mu_1 \ldots \mu_J} \Phi^{(\alpha)}_{\nu_1 \ldots \nu_J} Q_{\mu_1} \ldots Q_{\mu_J} P_{\nu_1} \ldots P_{\nu_J} = \\ &= C_J\ P_J(\cos\theta_t) \end{aligned} \tag{2.7}$$

where

$$C_j = \frac{(-)^J J!}{(2J-1)!!} |P|^J \; |Q|^J \quad ,$$

$$\cos\theta_t = \frac{(P\,Q)}{|P||Q|} = \frac{2\nu}{4m^2-t} = z \; .$$

P_J is a Legendre polynomial, m is the common π and ρ mass. The contribution of diagram 2.2 to T is given by

$$T^{(J)} = g^{(J)}\{[f_1^{(J)}(\varepsilon_1 \varepsilon_2) + f_2^{(J)}(\varepsilon_1 Q)(\varepsilon_2 Q)]\; F_J + \\ + f_3^{(J)}[\varepsilon_{1\mu}(\varepsilon_2 Q) + \varepsilon_{2\mu}(\varepsilon_1 Q)]\; \frac{\partial F_J}{\partial Q_\mu} + \\ + f_4^{(J)} \varepsilon_{1\mu}\; \varepsilon_{2\nu} \; \frac{\partial^2 F}{\partial Q_\mu \; \partial Q_\nu}\} \; \frac{1}{t-m^2} \quad , \tag{2.8}$$

with

$$\frac{\partial F_J}{\partial Q_\mu} = \frac{C_J}{Q^2} \; [P_\mu \; P_J'(z) - Q_\mu P_{J-1}'(z)] \quad , \tag{2.9}$$

$$\frac{\partial^2 F_J}{\partial Q_\mu \partial Q_\nu} = \frac{C_J}{Q^4} \; \{Q_\mu Q_\nu P_{J-2}'' - (P_\mu Q_\nu + P_\nu Q_\mu) P_{J-1}'' + P_\mu P_\nu P_J'' - \\ - Q^2[\delta_{\mu\nu} - \frac{\Delta_\mu \Delta_\nu}{m^2}] \; P_{J-1}'\} \; . \tag{2.10}$$

In the expression of the second derivative, care must be taken that the tensor $\partial^2 F/(\partial Q_\mu \partial Q_\nu)$ has vanishing components along the direction of the propagating four-vector Δ .

We finally obtain the contributions to the amplitudes:

$$\mathrm{Im}\; A^{(J)} = C_J g^{(J)} \; \frac{f_4^{(J)}}{Q^4} \; P_J''(z) \; \delta(t - m^2) \quad ,$$

$$\text{Im } B^{(J)} = C_J g^{(J)} \left[\frac{f_3^{(J)}}{Q^2} P_J'(z) - \frac{f_4^{(J)}}{2Q^4} P_{J-1}''(z)\right]\delta(t-m^2),$$

$$\text{Im } C_1^{(J)} = C_J g^{(J)}\{-f_2^{(J)} P_J(z) - \frac{f_3^{(J)}}{Q^2}[P_J'(z) + P_{J-1}'(z)] +$$

$$+ \frac{f_4^{(J)}}{4Q^4}[P_{J-2}''(z) + 2P_{J-1}''(z) + 2P_J''(z) - \frac{Q^2}{m^2}P_{J-1}'(z)]\}\times$$

$$\times\ \delta(t - m^2)\ ,$$

$$\text{Im } C_2^{(J)} = C_J g^{(J)}[f_1^{(J)} P_J(z) - \frac{f_4^{(J)}}{Q^2} P_{J-1}'(z)]\delta(t - m^2)\ . \qquad (2.11)$$

Now, the dependence upon the variable ν at fixed t is entirely contained in the Legendre polynomials, and we obtain that, for large ν

$$A \sim \nu^{J-2},$$
$$B \sim \nu^{J-1},$$
$$C_{1,2} \sim \nu^{J}. \qquad (2.12)$$

By reggeisation, $J \to \alpha(t)$, the trajectory value, and we obtain the high energy behaviour

$$A \sim \nu^{\alpha-2}\ ,$$
$$B \sim \nu^{\alpha-1}\ ,$$
$$C_{1,2} \sim \nu^{\alpha}\ .$$

We therefore see that some among the ρ - π scattering amplitudes show a more convergent asymptotic behaviour than the corresponding one for scalar particles. Therefore, if $\alpha < 1$ (as it is in charge exchange, where the ρ trajectory is dominant), we have the sum rule

$$\int \mathrm{Im}\ A^{(1)}(\nu,t)d\nu = 0 \qquad (2.13)$$

(the upperscript 1 means isospin T = 1 in the t channel). The result (2.12) is evident from an elementary point of view. The exchanged particle has J indices and any invariant obtained by contracting a vector (momentum or polarization) in the upper vertex with one in the lower vertex is obtained by use of one of those indices. So the maximum power in $\nu = \frac{1}{2}(P\ Q)$ is J. The products $(P\ \varepsilon_i)$ require one index, so that for $(P\varepsilon_1)(P\varepsilon_2)$ and $(P\varepsilon_1)(Q\varepsilon_2) + (P\varepsilon_2)(Q\varepsilon_1)$ respectively J-2 and J-1 indices are available to form powers of ν , which is just the result (2.12).

Let us note that (2.13) depends on very general requirements about analyticity in the angular momentum; the sum rule (2.13) holds in presence of any angular momentum pole or cut, provided that the largest value of α is smaller than one.

It may be of interest to note that the different high energy asymptotic behaviours can be understood from high energy unitarity requirements alone. Let us start with a very heuristic illustration of the Froissart high energy bound for a scalar particle amplitude A(s,t). We use the optical theorem at high energy, together with the condition that the total cross section be larger than the elastic one:

$$\mathrm{Im}\ A(s,o) > \frac{\text{const.}}{\sqrt{s(s-4m^2)}} \int_{-s+4m^2}^{o} |A(s,t)|^2\ dt\ . \qquad (2.14)$$

Let us now assume a constant shape of the diffraction peak: A(s,t) = A(s,o) f(t), with

$$\int_{-\infty}^{o} dt|f(t)|^2 < \infty \quad .$$

Then

$$\mathrm{Im}\ A(s,o) > \frac{\text{const.}}{s}\ |A(s,o)|^2\quad .$$

Now this demands that

$$|A(s,o)| < \text{const.}s \qquad (2.15)$$

Of course the condition of constant diffraction peak could be dropped by allowing slow logarithmic variation (like in the case of moving Regge poles); this would involve extra logarithmic terms in the high energy bound.

We now apply the same kind of argument to the ρ-π scattering. Let us use the "orthogonal" decomposition of the invariant amplitude T:

$$T = \alpha I_\alpha + \beta I_\beta + \gamma I_\gamma + \delta I_\delta \ , \qquad (2.16)$$

where

$$I_\alpha = (\varepsilon_1 P')(\varepsilon_2 P') \ ,$$

$$I_\beta = (\varepsilon_1 P')(\varepsilon_2 Q) + (\varepsilon_2 P')(\varepsilon_1 Q) \ ,$$

$$I_\gamma = (\varepsilon_1 Q)(\varepsilon_2 Q) \ ,$$

$$I_\delta = (\varepsilon_1 N)(\varepsilon_2 N) \ ,$$

$$P'_\mu = P_\mu - \frac{2\nu}{Q^2} Q_\mu \ ,$$

$$N_\mu = \varepsilon_{\mu\nu\rho\sigma} P_\nu Q_\rho \Delta_\sigma \ . \qquad (2.17)$$

The amplitudes α, β, γ, δ are connected to A, B, C_1, C_2 by the relations

$$A = \alpha + \frac{1}{4} Q^2 \Delta^2 \delta,$$

$$B = -\frac{4\nu}{Q^2} \alpha + \beta - \nu\Delta^2\delta \ ,$$

$$C_1 = \left(\frac{2\nu}{Q^2}\right)^2 \alpha - \frac{2\nu}{Q^2} \beta + \gamma + \left[\frac{P'^2}{4}(\Delta^2 - Q^2) + \frac{4\nu^2\Delta^2}{Q^2}\right]\delta ,$$

$$C_2 = -\frac{1}{4} P'^2 Q^2 \Delta^2 \delta .$$

We use again the optical theorem and ask that the total cross section be larger than each contribution of the four amplitudes α, β, γ, δ (because of the choice of invariants no interference term appears). We get then for large s

$$s^4 \int |\alpha(s,t)|^2 dt < C\sigma_t \; s^2 ,$$

$$s^2 \int Q^2 |\beta(s,t)|^2 t dt < C\sigma_t \; s^2 ,$$

$$\int Q^2 |\gamma(s,t)|^2 t^2 dt < C\sigma_t \; s^2 ,$$

$$s^4 \int |\delta(s,t)|^2 t^4 dt < C\sigma_t \; s^2 .$$

Again using the constant shape assumption, and expressing the result in terms of the old invariants, we have the asymptotic behaviour for large ν and fixed t

$$|A| < C \, s^{-1} ,$$

$$|B| < C ,$$

$$|C_{1,2}| < C \, s .$$

The discussion outlined above gives rise to a further question. The asymptotic behaviour of the "orthogonal decomposition" amplitudes is given by

$$|\alpha| < C \, s^{-1} ,$$

$$|\beta| < C ,$$

$$|\gamma| < C \, s ,$$

$$|\delta| < C\, s^{-1} \, . \tag{2.19}$$

From these formulae one could conclude that two amplitudes show a s^{-2} asymptotic factor with respect to the less convergent amplitude. Therefore, taking a Regge pole model and a ρ trajectory, we would have two sum rules of the kind (2.13), while we had got just one through the preceding discussion. The reason of such ambiguity lies in the fact that in order to derive sum rules from an asymptotic behaviour we pass through dispersion relations, and we have to make an appropriate choice of the invariants, such that dispersion relations hold, with the minimum number of subtractions required by asymptotic behaviour, and we have no kinematical singularity or zero in the variable s.

Now, if we start from the hypothesis that the amplitudes A, B, $C_{1,2}$ are devoid of kinematical singularities (and this is supported by the expressions (2.11), where it is seen that the ν dependence is entirely contained in the Legendre polynomials), then clearly the amplitude δ possesses a kinematical singularity at $P'^2 = 0$, i.e. when $\nu^2 = \frac{1}{4}P^2Q^2$. Such observation leads us to discuss the problem of the choice of invariants free from kinematical singularities in the energy, for which the asymptotic behaviour at fixed t can be determined, and possessing simple properties under crossing.

3. Invariants

The strong interaction superconvergence sum rules, and, as we will see, to a certain extent, the current algebra sum rules, start from the combination of good analytic properties and favourable asymptotic behaviour of scattering amplitudes. In addition, saturation of those sum rules requires the decomposition of the superconvergent amplitudes in the contribution of each partial wave (which could be conveniently approximated by isobars). We shall discuss here the problem of the choice of a convenient set of invariant amplitudes from this point of view.

The first requirement is the absence of kinematical singularities in the variable ν. Now, the problem of kinematical singularity free amplitudes has been understood in a general manner (see ref. [2]), although the procedure is somewhat cumbersome, so that in many simple cases one uses the indication of perturbation theory. However this criterion of choice is usually not sufficient to give a simple and general connection with the partial wave expansion and therefore the Regge pole analysis of the asymptotic behaviour is somewhat indirect.

In this respect the helicity amplitude formalism of Jacob and Wick has been suggested [3] as a possible candidate to play the dominant role in the theory of superconvergence. Indeed following Trueman let us consider the scattering shown in fig. 3.1,

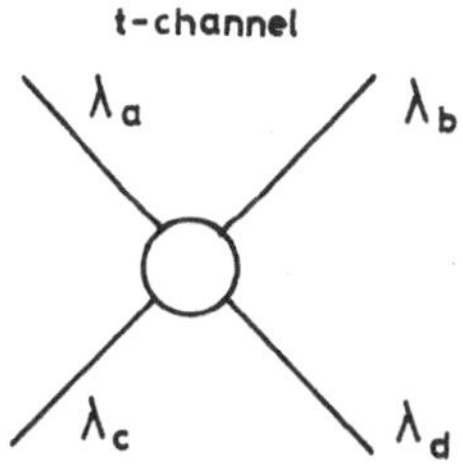

Fig. 3.1

Let $F_{\lambda_c\lambda_d,\lambda_a\lambda_b}(s,t)$ be a helicity amplitude for the t-channel reaction. Every term in the partial wave expansion of F contains the factor

$$f_{\lambda,\mu}(\theta_t) = (\cos\theta_t/2)^{|\lambda+\mu|}\ (\sin\theta_t/2)^{|\lambda-\mu|}$$

where

$$\mu = \lambda_c - \lambda_d \; , \quad \lambda = \lambda_a - \lambda_b \quad ,$$

and θ_t is the scattering angle in the t channel. Defining

$$A_{\lambda_c\lambda_d,\lambda_a\lambda_b} = \frac{F_{\lambda_c\lambda_d,\lambda_a\lambda_b}}{f_{\lambda,\mu}} \; ,$$

the function A is free from kinematical singularities in the energy.

Since, for large s and fixed t

$$\cos\theta_t \sim C\, s$$

we see for large s

$$|A_{\lambda_c\lambda_d,\lambda_a\lambda_b}| \sim C\, \frac{|F_{\lambda_c\lambda_d,\lambda_a\lambda_b}|}{s^{n(\lambda,\mu)}} \qquad ,$$

where $n(\lambda,\mu)$ is the maximum between λ and μ.

The bound on F for large t must be then established, connecting it to the helicity amplitudes in the crossed s channel $G_{\mu_b\mu_d,\mu_a\mu_c}(s,t)$ through the crossing relation

$$F_{\lambda_c\lambda_d,\lambda_a\lambda_b} = \Sigma_{\mu_i} (-1)^n \; d_{\mu_a\lambda_a}(\psi_a) d_{\mu_b\lambda_b}(\psi_b) d_{\mu_c\lambda_c}(\psi_c) \times$$

$$\times\, d_{\mu_d\lambda_d}(\psi_d)\; G_{\mu_b\mu_d,\mu_a\mu_c}(s,t) \quad . \qquad (3.2)$$

The factor $s^{-n(\lambda,\mu)}$ in (3.1) is the responsible for better convergence of some amplitudes, since all F's behave in the same way for large s, as it comes from (3.2).

However for practical use the helicity amplitudes lead to some difficulty, since crossing properties of those amplitudes are involved ((3.2)) so that in the utilization of superconvergence sum rules very strong cancellations between left and right hand cuts are expected in the last stage of calculations with one dimensional dispersion relations.

We wish to discuss here a procedure to write down an invariant decomposition, which might prove advantageous for the treatment of superconvergence [4]. The main advantage lies in the simple connection with partial wave amplitudes on one side and with the kinematical singularity free amplitudes on the other.

Let us consider a scattering process of particles and/or currents with the following kinematics shown in fig. 3.2

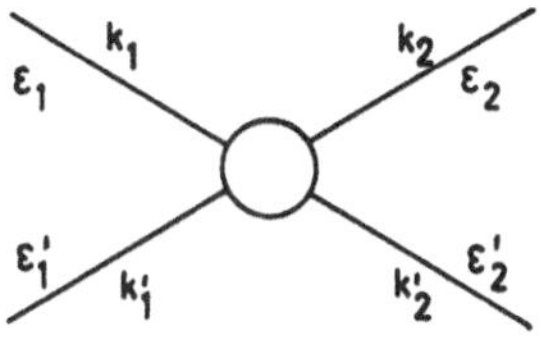

Fig. 3.2

$$
\begin{aligned}
k_i^2 &= m_i^2 \; ; \; k'^2_i = m'^2_i \\
P &= k'_2 - k_2 \\
K &= k_1 + k'_1 = k_2 + k'_2 \\
t &= K^2 \\
s &= (k_1 - k_2)^2 \\
u &= (k_1 - k'_2)^2
\end{aligned}
\qquad (3.3)
$$

$$
\begin{aligned}
\nu &= \tfrac{1}{2}(s - u) \\
z &= \cos\theta_t = \vec{k}_1 \cdot \vec{k}_2 / |\vec{k}_1||\vec{k}_2| = \alpha(t)\nu + \beta(t)
\end{aligned}
\qquad (3.4)
$$

$$
\alpha(t) = 2t\{[t-(m_1-m'_1)^2][t-(m_1+m'_1)^2][t-(m_2-m'_2)^2] \times
$$

$$
\times [t-(m_2+m'_2)^2]\}^{-1/2} \quad , \qquad (3.4')
$$

$$
\beta(t) = \frac{\alpha(t)}{2t}(m_1^2-m'^2_1)(m_2^2-m'^2_2) \; .
$$

In fig. 3.2 the ε's are polarization tensors describing the spins (ε stands for $\varepsilon_{\mu_1 \cdots \mu_n}$ if the spin is n) or γ matrices. We have at left the incoming states (all variables referring to it are labelled by "1"), at right the outgoing state labelled by "2". The connection between the two sets of variable "1" and "2" is given by

$$K = k_1 + k'_1 = k_2 + k'_2$$

and by conservation of the total angular momentum J.

We may identify the angular momentum J contribution to the various amplitudes with the help of the following observations. Let us consider the behaviour of the amplitude under rotation in the center of mass system (that is, those transformations which keep K constant). Let us indicate by R_1^{α} the rotation (characterized by the Euler angles α) of the kinematical variables 1 and by R_2^{β} the rotation (characterized by β) of the variables 2. Overall rotational invariance requires that the amplitude is invariant under the product $G^{\alpha\alpha} = R_1^{\alpha} R_2^{\alpha}$ of the same rotation performed both on the variables 1 and 2.

Now it is convenient to introduce the larger group of transformation

$$G^{\alpha\beta} = R_1^{\alpha} R_2^{\beta}$$

where rotations α and β are not correlated.

It is clear that, while the scattering matrix T is invariant under operations of the type $G^{\alpha\alpha}$, it is certainly not so in general for separate rotation R_1^{α} (or R_2^{β}) of the initial (or final) variables, except when the initial state is a scalar under rotations, and one realizes immediately that an amplitude corresponding to a fixed J transforms as a J representation under R_1 (or R_2) operation.

These considerations give a simple general prescription about the most convenient way of combining momenta and polarizations to form invariant factors. Indeed, if we combine the variables "1" and "2" to obtain factors, embodying the complete polarization dependence, which are invariants under R_1, or R_2 operations separately, then the coefficient of these factors, in an expansion of the amplitude T, enjoys the property that a spin J contribution transforms like a J representation under separate R_1, R_2 operations. Such factorization is attained by combining separately quantities of type "1" or "2".

In order however to get the right number of invariants, we must use, besides ε, γ_μ, k_μ, k'_μ also the gradient $\partial/\partial k_\mu$, being plain that such gradient must be performed at constant K_μ so that only one independent gradient exists for state "1", and one for state "2".

Indeed a gradient, e.g. $\partial/\partial k_{1\mu}$, acting upon a scalar amplitude $\Phi(t, \cos\theta_t)$, at constant K_μ and therefore at constant t, will reproduce a term proportional to $(k'_2 - k_2)_\mu$, since

$$\cos\theta_t = \alpha(t)\,\nu + \beta\,(t)$$

and

$$\nu = \frac{1}{2}[k_2^2 - k_2'^2 + 2(k_1 \cdot (k'_2 - k_2))] \quad .$$

In performing the $\partial/\partial k_\mu$ operation one has to be careful to consider the dependence on all variables.

Indeed if one performs this derivative the "masses" associated with the external lines do indeed vary, and one has to consider this mass dependence of the amplitudes. One can of course perform a differentiation in which the masses are kept constant. In this way, however, writing $d\Phi = \Phi_\mu dk_\mu$ the differentials dk_μ are not independent but are related by the orthogonality conditions $k_\mu dk_\mu = K_\mu dk_\mu = 0$. Therefore in this case the invariants coming from the gradients must be obtained by saturating them with polarization vectors or tensors properly orthogonalized to k_μ and K_μ. The two procedures are equivalent.

Keeping the masses constant is preferable in strong interaction, while in current algebra it may be convenient to perform the derivatives with respect to the "masses" of the currents.

Scalar products between left and right polarization tensors are obtained through the antisymmetric $\varepsilon_{\mu\nu\rho\sigma}$ Indeed in the development of $\varepsilon_{\mu\nu\rho\sigma}\varepsilon_{1\mu}\varepsilon'_{1\nu}k_{1\rho}k'_{1\sigma}\varepsilon_{\alpha\beta\gamma\delta}\varepsilon_{2\alpha}\varepsilon'_{2\beta}k_{2\gamma}k'_{2\delta}$ terms as $(\varepsilon_1\varepsilon_2)$ appear.

We may write therefore the decomposition of the invariant amplitude T as *

$$T = \sum_{m,n} \alpha_m^{(1)} \alpha_n^{(2)} \Phi_{mn} , \tag{3.5}$$

where the $\alpha_m^{(1)}$ are scalar combinations of the polarization tensors (and γ_μ's if half integer-spins are present) with vectors and gradient of side "1", and correspondingly $\alpha_n^{(2)}$ are scalars formed from side "2" quantities. Φ_{mn} is a pure Lorentz invariant function of the momenta. Of course in an actual scattering, symmetry principles such as parity or time reversal, and supplementary conditions on the polarization tensors, may further limit the number of independent amplitudes.

The considerations developed above imply that, for total angular momentum J of the initial state, the Φ_{mn}, being scalars under simultaneous rotations $R_1^\alpha R_2^\alpha$, behave like representations of dimension 2J+1 under R_1 or R_2 separately. Therefore a spin J contribution to Φ_{mn} takes the form $\Phi_{mn}^J(t)\, P_J(z)$, where $t = K^2$ and $z=\cos\theta_t$.

The decomposition (3.5) tells us that, in the gener-

* We will limit ourselves to the case of integer total spin in the t channel. The extension to half integer total spin is not difficult.

al case, writing

$$\Phi_{mn}(t,z) = \sum_J \Phi^J_{mn}(t) \, P_J(z) \quad , \tag{3.6}$$

the functions Φ^J_{mn} describe contributions from intermediate states of total angular momentum J. The existence of various functions for fixed J is of course related to the different ways in which initial and final spins and orbital momenta can combine to given total angular momentum J.

The procedure outlined here has many important advantages. First of all the amplitudes $\Phi_{mn}(t,z)$ are, at fixed t, free of kinematical singularities in the variable z.

Secondly, the simple expansion (3.6), reminiscent of the theory of spinless particles, allows a direct application of the Regge procedure of analytic continuation in J.

Finally, it is easy to obtain from eqs. (3.5), (3.6) a general isobaric model by simply setting

$$\mathrm{Im}\ \Phi^J_{mn}(t) = f^{(J)}_m \ f^{(J)}_n \ \delta(t - m^2_J) \tag{3.7}$$

where m_J is the isobar mass and the $f^{(J)}_m$'s are the appropriate coupling constants.

Another interesting consequence of our technique can be drawn when from the original T we extract other amplitudes by applying contraction operations, as for instance substitutions $\varepsilon_\mu \rightarrow k_\mu$ or $\varepsilon_\mu \varepsilon_\nu \rightarrow g_{\mu\nu}$.

Let us investigate in particular the case when such contractions do not mix quantities "1" with quantities "2", as it is in the case of current algebra, where the substitution $\varepsilon_\mu \rightarrow k_\mu$ is very useful. Such operations are then invariant under separate R_1, R_2 rotations, and the considerations previously developed hold for the quantities generated in this way. So, if the initial state possesses spin J, a contracted quantity U of the kind above described behaves under rotations as a representation of spin J. Expanding U in an equivalent form to (3.5)

$$U = \sum_{m,n} \beta_m^{(1)} \beta_n^{(2)} \psi_{mn}(t,z)$$

with

$$\psi_{mn}(t,z) = \sum_J \psi_{mn}^J(t) P_J(z)$$

the quantity ψ_{mn}^J corresponds to states of total angular momentum J, and therefore can be expressed as a linear combination with coefficients depending on t but not on ν, of the Φ_{mn}^J, the analogous terms in the decomposition of the amplitude T from which U was derived.

Let us go back to the form of the amplitude T as expressed in eq. (3.5). We see that the amplitudes Φ_{mn} are determined only apart from a polynomial in ν whose degree depends on the number of derivative operators appearing in the corresponding $\alpha_m^{(1)} \alpha_n^{(2)}$. It is thus convenient to perform the differentiation and to obtain a new set of invariants which are linear combinations (with coefficients free from kinematical singularities in ν) of the Φ_{mn} and their derivatives with respect to ν.

Those invariant functions will therefore have no kinematical singularities at fixed t.

On the other side, as it will be seen more clearly in the following examples, our procedure allows to obtain in a simple way the Regge asymptotic behaviours.

Let us illustrate our procedure on some examples, referring to the kinematics defined in fig. 3.2.

In the first example we shall assume lines 1 and 1' as representing vector currents while lines 2 and 2' correspond to scalar or pseudoscalar objects (particles or divergences). This example will reveal useful when discussing current algebra. We consider then a tensor amplitude $T_{\mu\nu}$ defined by $T = \varepsilon_{1\mu}\varepsilon'_{1\nu}T_{\mu\nu}$. Let us expand such amplitude following our procedure. The side two invariants $\alpha_n^{(2)}$ reduce to unity. We shall then construct suitable rank two tensors using the vectors $k_{1\mu}$, $k'_{1\nu}$, the gradient $\partial/\partial k_{1\mu}$ and $g_{\mu\nu}$. When introducing the gradient we must exercise some care since, as previously pointed out, in this case we perform the derivative at fixed masses and fixed K_μ. We achieve this through the definition of a projection tensor $I_{\mu\nu}$ which projects every vector into a subspace orthogonal to k_1 and k'_1. Such tensor takes the form

$$I_{\mu\nu} = \delta_{\mu\nu} - \Delta_{\mu\nu}$$

where

$$\Delta_{\mu\nu} = - \left\{k_1^2 k'_{1\mu} k'_{1\nu} + k'^2_1 k_{1\mu} k_{1\nu} - (k_1 k'_1)(k_{1\mu} k'_{1\nu} + k'_{1\mu} k_{1\nu})\right\} \times$$

$$\times \frac{1}{\gamma(t)}$$

and

$$\gamma(t) = (k_1 k'_1)^2 - k_1^2 k'^2_1 = \frac{1}{4}\left[t-(m_1-m'_1)^2\right]\left[t-(m_1+m'_1)^2\right] .$$

It is immediate to recognize that $I_{\mu\nu} = I_{\nu\mu}$ and $I_{\mu\nu} k_{1\nu} = I_{\mu\nu} k'_{1\nu} = 0$; so $I_{\mu\nu}$ projects every vector in the subspace orthogonal to k_1 and k'_1 .

In our decomposition we shall then always introduce $I_{\mu\nu}(\partial/\partial k_{1\nu})$ in place of the gradient $\partial/\partial k_{1\mu}$. As a consequence the derivatives will be performed at constant k_1^2, k'^2_1, $(k_1 k'_1)$ and t.

Our decomposition of $T_{\mu\nu}$ is then

$$T_{\mu\nu} = \sum_{m=1}^{10} \alpha_{\mu\nu}^{(m)} \Phi_m(\nu,t) \quad , \qquad (3.8)$$

where

$$\alpha_{\mu\nu}^{(1)} = I_{\mu\alpha} I_{\nu\beta} \frac{\partial}{\partial k_{1\alpha}} \frac{\partial}{\partial k_{1\beta}} \, , \qquad \alpha_{\mu\nu}^{(6)} = k_{1\mu} k_{1\nu} \, ,$$

$$\alpha_{\mu\nu}^{(2)} = k_{1\nu} d_\mu \, , \qquad \alpha_{\mu\nu}^{(7)} = k_{1\mu} k'_{1\nu} \, ,$$

$$\alpha_{\mu\nu}^{(3)} = k'_{1\nu} d_\mu \, , \qquad \alpha_{\mu\nu}^{(8)} = k'_{1\mu} k_{1\nu} \, ,$$

$$\alpha_{\mu\nu}^{(4)} = k_{1\mu} d_\nu \, , \qquad \alpha_{\mu\nu}^{(9)} = k'_{1\mu} k'_{1\nu} \, ,$$

$$\alpha_{\mu\nu}^{(5)} = k'_{1\mu} d_\nu \, , \qquad \alpha_{\mu\nu}^{(10)} = g_{\mu\nu} \, ,$$

$$d_\mu = I_{\mu\alpha} \frac{\partial}{\partial k_{1\alpha}} \, .$$

From the previous discussion it follows immediately that each invariant amplitude $\Phi_m(\nu,t)$ can be expanded in partial waves

$$\Phi_m(\nu,t) = \sum_J g_m^{(J)}(t) \; P_J(z)$$

where $g_m^{(J)}$ entails all contributions from intermediate states with angular momentum J.

We can now explicitly perform the derivatives in (3.8), taking into account that

$$d_\mu = I_{\mu\alpha} \; P_\alpha \frac{\partial}{\partial \nu} \, .$$

The resulting decomposition assumes then the form

$$T_{\mu\nu} = A_1 P_\mu P_\nu + A_2 P_\mu k_{1\nu} + A_3 P_\mu k'_{1\nu} + A_4 k_{1\mu} P_\nu + A_5 k'_{1\mu} P_\nu +$$
$$+ A_6 k_{1\mu} k_{1\nu} + A_7 k_{1\mu} k'_{1\nu} + A_8 k'_{1\mu} k_{1\nu} + A_9 k'_{1\mu} k'_{1\nu} + A_{10} g_{\mu\nu} , \qquad (3.9)$$

where

$$
\begin{aligned}
A_1 &= \Phi''_1 \\
A_2 &= \omega\Phi''_1 + \Phi'_2 \\
A_3 &= \omega'\Phi''_1 + \Phi'_3 \\
A_4 &= \omega\Phi''_1 + \Phi'_4 \\
A_5 &= \omega'\Phi''_1 + \Phi'_5 \\
A_6 &= \omega^2\Phi''_1 + \omega(\Phi'_2 + \Phi'_4) + \Phi_6 \\
A_7 &= \omega\omega'\Phi''_1 + \omega\Phi'_3 + \omega'\Phi'_4 + \Phi_7 \\
A_8 &= \omega\omega'\Phi''_1 + \omega'\Phi'_2 + \omega\Phi'_5 + \Phi_8 \\
A_9 &= \omega'^2\Phi''_1 + \omega'(\Phi'_3 + \Phi'_5) + \Phi_9 \\
A_{10} &= \Phi_{10}
\end{aligned}
\tag{3.10}
$$

and

$$
\omega(\nu,t) = \frac{1}{2\gamma(t)}\left[(t-u+u')\nu - \tfrac{1}{2}(t-u-3u')(v'-v)\right] \quad ,
$$

$$
\omega'(\nu,t) = -\frac{1}{2\gamma(t)}\left[(t-u'+u)\nu - \tfrac{1}{2}(t-u'-3u)(v-v')\right] \quad ,
$$

$$
u = k_1^2 \ , \ u' = k_1'^2 \ , \ v = k_2^2 \ , \ v' = k_2'^2 \quad . \tag{3.11}
$$

We notice that the invariant amplitudes $A_1 \ldots A_{10}$ are the ones spontaneously introduced in a perturbative expansion of $T_{\mu\nu}$. In this sense we shall often refer to the A_i as perturbative invariants.

As a second example we consider the case where the lines 1 and 1' represent vector mesons (say the ρ meson) and the lines 2 and 2' pseudoscalar mesons (say π). Then we must fulfil the supplementary conditions

$$
(\varepsilon_1 k_1) = (\varepsilon'_1 k'_1) = 0 \quad . \tag{3.12}
$$

The physical amplitude for this process, T, can be easily obtained from the previous one by contracting the expression (3.8) for $T_{\mu\nu}$ with $\varepsilon_{1\mu}\varepsilon'_{1\nu}$, and taking into account conditions (3.12) and time reversal to reduce to four the number of independent amplitudes in the decomposition of T. We have thus

$$T = (\varepsilon_1^T \frac{\partial}{\partial k_1})(\varepsilon_1'^T \frac{\partial}{\partial k_1})\,\psi_1 +$$

$$+ \{(\varepsilon' k_1)(\varepsilon_1^T \frac{\partial}{\partial k_1}) - (\varepsilon_1 k_1')(\varepsilon_1'^T \frac{\partial}{\partial k_1})\}\psi_2 +$$

$$+ (\varepsilon' k_1)(\varepsilon_1 k_1')\psi_3 + (\varepsilon_1 \varepsilon_1')\psi_4 \quad , \qquad (3.13)$$

where we have introduced transversal polarization vectors

$$\varepsilon_\mu^T = I_{\mu\nu}\,\varepsilon_\nu \;, \qquad \varepsilon_\mu'^T = I_{\mu\nu}\,\varepsilon_\nu' \;,$$

and

$$\Phi_1 \to \psi_1 \;,\; \Phi_8 \to \psi_3 \;,\; \Phi_{10} \to \psi_4$$

and

$$\Phi_2 \to \psi_2 \;,\; \Phi_5 \to -\psi_2$$

as imposed by time reversal invariance.

By explicitly carrying on the derivatives we find

$$T = (\varepsilon_1 P)(\varepsilon_1' P)A + \left[(\varepsilon_1 P)(\varepsilon_1' k_1) - (\varepsilon_1 k_1')(\varepsilon_1' P)\right] B -$$

$$- (\varepsilon_1 k_1')(\varepsilon_1' k_1) C_1 + (\varepsilon_1 \varepsilon_1')\, C_2 \quad , \qquad (3.14)$$

where

$$A = \psi_1'' \;, \qquad B = \omega\psi_1'' + \psi_2'$$

$$C_1 = \omega^2 \psi_1'' + 2\omega\, \psi_2' - \psi_3 \;, \quad C_2 = \psi_4 \qquad (3.15)$$

and

$$\omega = \frac{2\nu}{t - 4m_\rho^2} \quad . \tag{3.16}$$

As a final simple example we sketch the case where 1 and 1' are respectively nucleon and antinucleon, and 2 and 2' pion lines.

We write the physical amplitude as

$$T = \bar{u}(-k_1') \, [(\gamma_\mu^T \frac{\partial}{\partial k_{1\mu}})\Phi_1 + \Phi_2]u(k_1) \quad , \tag{3.17}$$

where

$$\gamma_\mu^T = I_{\mu\nu}\, \gamma_\nu \quad , \qquad (\gamma^T k_1) = (\gamma^T k_1') = 0 \quad .$$

The connection with the usual invariants A, B can be easily seen. Performing the derivatives we have

$$T = \bar{u}(-k_1')\{A + B(\gamma Q)\}u(k_1) \quad , \tag{3.18}$$

where, respecting the tradition, we have renamed $Q = k_2' - k_2$ the vector previously named P and

$$A = \frac{4m\nu}{t-4m^2}\,\Phi_1' + \Phi_2 \quad ,$$

$$B = \Phi_1' \quad . \tag{3.19}$$

As we can immediately realize from the examples presented above, it is an easy matter to derive, for a given process,the high ν behaviour of the perturbative invariants at fixed t. It is indeed enough to notice that a spin J state exchanged in the t channel produces a contribution to the amplitude Φ_{mn}, given by eq. (3.6), which for large ν behaves as $K(t)\,\nu^J$. As seen in the examples, the perturbative invariants are combinations of the Φ_{mn} and of their derivatives; therefore, the behaviour for large ν of a spin J contribution to a perturbative amplitude A is $K(t)\nu^{J-r}$, where r is the minimum order of derivative of the Φ_{mn}'s appearing in A.

By Reggeization $J \to \alpha(t)$, where $\alpha(t)$ is the corresponding Regge trajectory, and we obtain the result that a perturbative invariant A_i behaves for large ν and fixed t as *

$$A_i \sim K(t)\nu^{\alpha(t)-r_i} . \tag{3.20}$$

In particular for ρ-π scattering we obtain the behaviours

$$A \sim \nu^{\alpha-2} , \; B \sim \nu^{\alpha-1} , \; C_{1,2} \sim \nu^{\alpha} , \tag{3.21}$$

and for π-N scattering ,

$$A \sim \nu^{\alpha} , \qquad B \sim \nu^{\alpha-1} . \tag{3.22}$$

Let us now summarize the preceding discussion. We have seen that the so called "perturbative invariants" can be expressed in terms of the amplitudes Φ_{mn} , with coefficients which do not introduce kinematical singularities. The amplitudes have no kinematical singularity in ν , and their partial wave analysis and analytic continuation in J are straightforward.

This discussion shows also very clearly the meaning of superconvergence .

Indeed the better convergence of the spin flip amplitudes is directly connected with the derivatives with respect to ν, appearing in eqs. (3.10), (3.15) and (3.19) which relate the two sets of invariants.

---

* We remark that this procedure was essentially proposed in the paper by M. Gell-Mann, S.Frautschi and F. Zachariasen [5].

4. Applications of Superconvergence Sum Rules

We have seen how to establish asymptotic behaviours for the amplitudes relative to scattering of particles endowed with spin. From the "superconvergent" asymptotic behaviour we may establish sum rules by means of fixed t dispersion relations as said above. In general the following theorem holds:

If, for large ν

$$|f(\nu)| \sim \nu^{b} , \qquad b < -n - \varepsilon$$

and

$$f(\nu) = \frac{1}{\pi} \int_{-\infty}^{+\infty} \frac{\mathrm{Im}\ f(\nu')}{\nu'-\nu} d\nu' ,$$

then

$$\int \nu^{P}\ \mathrm{Im}\ f(\nu) d\nu = 0 , \qquad P = 0,\ldots,n-1 . \tag{4.1}$$

Proof of this theorem is given in App. 1.
Having thus established superconvergence sum rules in strong interactions, we are confronted with two problems. On one hand there is the problem of practical utilization of the sum rules, obtaining links between experimental quantities, and predictions; on the other hand, considering the present spectrum of resonant states, we are in presence of a number of sum rules increasing with the spins of the external particles, and we may ask how stringent is this set of sum rules, and what can be said about intermediate states; in brief, the general problem of saturation of all sum rules. We shall defer to a next chapter some remarks about such general problem; here we shall try to present a list of the applications of superconvergent sum rules, although such a review will be necessarily incomplete, since the

subject is very rapidly developing.

For the practical utilization of sum rules one may introduce as intermediate states, the known resonances. Let us consider as first example the case of ρ-π scattering [1], using the invariant decomposition (2.4). We have the asymptotic behaviours

$$A(\nu,t) \sim \nu^{\alpha-2} \quad , \; B(\nu,t) \sim \nu^{\alpha-1} \quad , \; C_{1,2}(\nu,t) \sim \nu^{\alpha} . \qquad (4.2)$$

Let us consider first the isospin T = 1 state in the t channel. The ρ trajectory is then dominant, with a trajectory value $\alpha(o) \simeq 0.5$ obtained from experimental data. Under such conditions we have the sum rule

$$\int_0^\infty \mathrm{Im}\, A^{(1)}(\nu,t)d\nu = 0 \quad , \qquad (4.3)$$

having used the crossing property $\mathrm{Im}\, A^{(1)}(-\nu,t) = \mathrm{Im}\, A^{(1)}(\nu,t)$.

(4.3) can be put under the form of an infinite set of sum rules for the imaginary part of the forward amplitude and for its derivatives with respect to t.

It is evident that, the higher is the order of derivatives the higher must become the importance of high angular momentum states. Therefore we begin by considering the forward direction sum rule

$$\int \mathrm{Im}\, A^{(1)}(\nu,o)d\nu = 0 \; . \qquad (4.4)$$

We now try to saturate (4.4) by introducing the lowest known states, and therefore we introduce the π and ω contributions (the ϕ is experimentally known not to be practically coupled to three pions). We proceed to calculate the contributions under the hypothesis of narrow resonance width. Notice that, because superconvergence sum rules are written at fixed t, such hypothesis is perfectly consistent and one never runs into troubles, as for instance in a partial wave dispersion relation, where particle poles give rise to a continuum in the left hand cut. Here, in the limit of zero width, we obtain algebraic equations relating masses and coupling constants without any further approximation .

We calculate then the contribution to the sum rule (4.4) due to the diagram

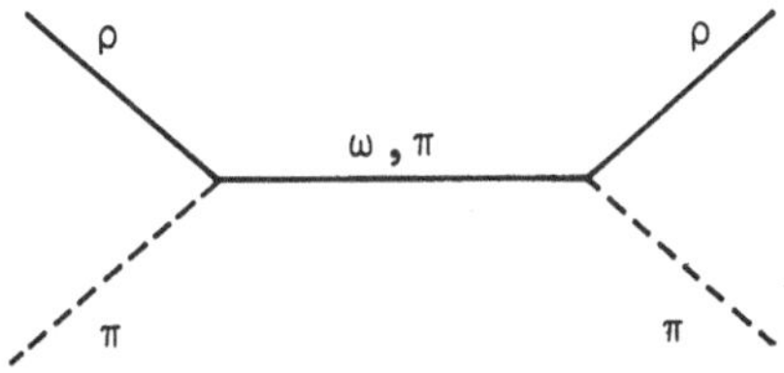

where the intermediate state is an ω or a π meson. The couplings are used

$$\rho\pi\pi)\quad g_{\rho\pi\pi}\ \varepsilon_{ijk}\ \rho^i_\mu\ \pi^j\ \partial_\mu\ \pi^k\ ,$$

$$\omega\rho\pi)\quad i\ g_{\rho\omega\pi}\ \varepsilon_{\alpha\beta\gamma\delta}\ \partial_\alpha\ \omega_\beta\ \partial_\gamma\ \rho^i_\delta\ \pi^i\ .$$

(4.4) then yields the following relation among the two coupling constants involved:

$$4g^2_{\rho\pi\pi} - m^2_\rho\ g^2_{\omega\rho\pi} = 0 \qquad (4.5)$$

(m_ρ is the ρ mass). Eqs. (4.4) and (4.5) were first obtained through current algebra by Fubini and Segré [6].

Taking $(g^2_{\rho\pi\pi})/4\pi = 3$, corresponding to a width of 150 MeV, through the Gell-Mann, Sharp and Wagner [7] model one gets a partial width $\Gamma(\omega \to 3\pi) \sim 8$ MeV, against an experimental value of (10 ± 2). The agreement is not too bad, considering that several approximations have been made.

Other forward sum rules can be deduced for double charge exchange amplitudes if we assume that no isospin T = 2 Regge pole exists with $\alpha_{++}(o) > 0$. Then $B^{(2)}(\nu,t)$ is also superconvergent, with the right crossing properties to yield a non trivial sum rule

$$\int_0^\infty \mathrm{Im}\, B^{(2)}(\nu,t)d\nu = 0 \quad . \tag{4.6}$$

From $A^{(2)}$ we can derive two sum rules, one identically satisfied by crossing (the analogous of (4.4)), the other being

$$\int_0^\infty \nu\, \mathrm{Im}\, A^{(2)}(\nu,t)d\nu = 0 \quad . \tag{4.7}$$

The system of algebraic equations, derived from (4.4), (4.6) and (4.7) in the forward direction, when saturated with π, ω, ϕ, is quite unsatisfactory, leading to

$$g^2_{\omega\rho\pi} = g^2_{\phi\rho\pi} = g^2_{\rho\pi\pi} = 0 \ .$$

P. M. Frampton and J. C. Taylor [8] have investigated the saturation of (4.4), (4.6) and (4.7) in the forward direction, and of their first derivatives with respect to t in the forward direction, introducing intermediate π, ω, A_1 and A_2 states in the approximation of zero width.

The couplings $(A_1\rho\pi)$ and $(A_2\rho\pi)$ are given by

$$g_{\rho^+A_1^o\pi}\ [(\varepsilon^{(\rho)}\varepsilon^{(A_1)}) + \frac{C}{m_\rho m_{A_1}}(\varepsilon^{(\rho)}p^{(A_1)})(\varepsilon^{(A_1)}p^{(\rho)})],$$

$$g_{\rho^+A_2^o\pi}\ \varepsilon_\alpha^{(\rho)}\varepsilon_{\beta\gamma}^{(A_2)}p_\gamma^{(\rho)}p_\delta^{(\rho)}p_\varepsilon^{(A_2)}\varepsilon_{\alpha\beta\delta\varepsilon} \quad .$$

For the $\rho A_1\pi$ constants, the values $(g_{\rho^+A_1^o\pi}) \simeq 1.9$ BeV and $C \simeq 10$ are used, which account for a with of $A_1 \to \rho\pi$ of 130 MeV, according to the formula [9]

$$\Gamma(A_1 \to \rho\pi) = \frac{P_\pi}{12\pi m_{A_1}^2} g^2_{\rho^+ A_1^o \pi} (3+0.2\ C + 0.01C^2) \ .$$

For $|g_{\rho A_2 \pi}|$ the value 20 GeV^{-2} is used, obtained from the decay width

$$\Gamma(A_2 \to \rho\pi) = \frac{P_\pi^5}{80\pi} g^2_{\rho A_2 \pi} \quad .$$

The authors find a nice saturation of the forward sum rules; however two out of the three sum rules for the first t derivatives are not so satisfying and require further contributions.

Let us pass to consider pseudoscalar meson baryon scattering. The invariant amplitude is decomposed as usually (see (3.18)) and we have the asymptotic behaviour of the two invariant functions (see 3.22) for $\nu \to \infty$, fixed t:

$$A(\nu,t) \sim \nu^{\alpha(t)} \ , \quad B(\nu,t) \sim \nu^{\alpha(t)-1} \quad . \tag{4.8}$$

Let us examine π-N scattering:

In the t channel ($\pi\pi \to N\bar{N}$) the isospin state T = 1 is dominated by the ρ trajectory. Now, $\alpha_\rho(o) \simeq 0.5$ and no amplitude is superconvergent.

The situation changes if we consider pseudoscalar meson-baryon in a SU(3) scheme.

From the experimental evidence indicating no low mass meson state with T or Y = 2 (which would be contained in a 27-plet in SU(3)) one can again rather reasonably assume that

$$\alpha_{27}(o) < 0 \quad . \tag{4.9}$$

Consequently, since $B^{(27)}(\nu,t)$ has the right crossing symmetry, the sum rule can be established

$$\int_0^\infty Im\ B^{(27)}(\nu,t)d\nu = 0 \quad . \tag{4.10}$$

Saturation of this sum rule in the forward direction has been discussed introducing as intermediate states the known resonances by Altarelli-Buccella-Gatto, Babu-Gilman-Suzuki, and Sakita-Wali [10]. The saturation is very sensitive to the D/F ratio of the baryon octet and is achieved with $(\frac{1}{2})^+$ and $(\frac{3}{2})^+$ for a value $\frac{D}{D+F} = 0.58$. With D/F = 1.5 and equal masses for the baryon octet and $(\frac{3}{2})^+$ decuplet the relation of SU(6) is obtained between g_π and the decuplet width. This reflects the fact that putting some consequences of a symmetry group into superconvergence equations, the results of that group are solutions of the equations, since obviously superconvergence cannot discriminate among internal symmetry groups [11].

One can of course go beyond the forward sum rule. Altarelli, Buccella and Gatto have considered the sum rules for the first and second derivative with respect to t in the forward direction, and obtained a reasonable agreement by including the contribution of higher resonant states. On the other hand Sakita and Wali have made the interesting observation that, for intermediate states of low nearby masses, the sum rule should be saturated separately for each set of resonances with the same orbital angular momentum ℓ , in order that the sum rule be satisfied for a range of values in t.*

Accordingly, relations are given among the widhts of resonances; the one which can be experimentally tested is

$$\frac{\Gamma_\Delta R_\Delta(1238)}{6g_\pi^2} = \frac{\Gamma_\Delta^{el} R_\Delta\ (1924)}{\Gamma_N^{el} R_N(1688)} = -\frac{8}{3}(2f_\alpha^2 + 2f_\alpha - 1) ,$$

where the R are kinematical coefficients, $\Delta(1924)$ is a $(\frac{7}{2})^+$, and N(1688) a $(\frac{5}{2})^+$. With experimental data

$$\frac{\Gamma_\Delta R_\Delta(1238)}{6g_\pi^2} = 0.51 \quad .$$

---

* We will see the reason in the next chapter.

while

$$\frac{\Gamma_{\Delta}^{el} R_{\Delta}(1924)}{\Gamma_{N}^{el} R_{N}(1688)} = 0.49 \sim 0.69 \quad .$$

Thus the agreement with experiments is fairly good.

D'Auria and de Alfaro [12] have discussed superconvergence sum rules in nucleon-nucleon scattering. The high t behaviour of fixed energy in the nucleon number two channel is examined, and it is concluded that, for the kinematical singularity free "β-decay" invariants, two sum rules can be written for vector and tensor amplitudes with isospin T = 1 in the fixed energy channel.

These sum rules, in the forward direction in the NN channel, are then saturated through the exchange of particles and resonances in the baryon number zero channels. The sum rules transform into relations between coupling constants. From the point of view of nucleon-nucleon interaction, these relations can be interpreted as conditions on the "potential", since the exchange of each single state in the t channel will yield a contribution which, at large t, will be less bounded than what required, and the relations among coupling constants deducted from the sum rules insure that the "potential" will be less singular than each single contribution.

The main contribution to the sum rules come from the π and the magnetic ρ coupling, and they almost cancel.

It is amusing to note that, limiting to those contributions, the relation obtained between g_{π} and $g_{2\rho}$ in the limit of infinite nucleon mass in the same as it can be obtained calculating the nucleon-nucleon potential in a static model with π and ρ, and then asking that the most singular terms at small distances in the potential cancel.

Other couplings (ω,ρ electric) give smaller contributions which tend again to cancel.

Other interesting investigations have been carried on, which cannot be reviewed in detail here, about electro - and photoproduction of pions on baryons, $\pi+N\to\pi+N^*$, etc. We give in note a short list of papers of which we are aware [13].

Concluding this short summary of some applications of superconvergence, we can say that we are encouraged, by the success of the practical utilizations, to investigate further the general implications of the sum rules. Because of the scarcity of knowledge of experimental data on coupling constants we are unfortunately limited to utilize a few sum rules with the lowest known states. It will be very interesting to develop further the program of practical utilization when more data will be available.

In the next chapter we shall try to develop some considerations which may shed some light on the general problem of saturation, discussing the number of angular momentum states needed, and giving, in a model, a general set of intermediate states and couplings fulfilling superconvergence requirements.

5. Considerations on Saturation of Superconvergence Sum Rules

We have seen in the last section some promising utilizations of superconvergence sum rules obtained by saturation with the lowest isobars. In this way the equations become pure algebraic relations involving masses and coupling constants. The main justification for such a procedure is phenomenological.

We know since long a situation where a few isobaric states give a reasonable description of phenomena in a certain energy range: low energy pion physics can be mainly explained in terms of nucleon and N^* contributions, where the (πNN) and (πNN^*) coupling constants are connected by a relation of superconvergence type.

Let us begin by discussing the problem of saturation of superconvergence sum rules by means of a set of "particles" of equal mass. Let us investigate a scattering process of particle endowed with spin, and let $A_i(\nu,t)$, $\bar{A}_i(\nu,t)$ be our invariant functions describing the process. The functions $\bar{A}_i$ are the superconvergent amplitudes; i.e. we have requirements of the kind

$$\int \mathrm{Im}\ \bar{A}_i(\nu,t)d\nu = 0 \ . \tag{5.1}$$

Saturating with a set of isobars of equal mass, we have

$$\mathrm{Im}\ \bar{A}_i(s,t) = \bar{a}_i(t)\ \delta(s-s_o) \quad , \tag{5.2}$$

where s_o is the common $(\mathrm{mass})^2$. Then, (5.1) requires $\bar{a}_i(t) = 0$. Now, we may expand $\bar{a}_i(t)$ in partial waves,

$$\bar{a}(t) = \Sigma_J\ C_J\ P_J(1 + \frac{t}{2k_o^2}) \quad ,$$

and we end up with the set of equations

$$C_J = 0 \quad , \qquad \text{for each } J \quad . \tag{5.3}$$

Since the external particles possess spin, the contribution to each C_J will be due to isobars of nearby angular momentum, and the numbers of different angular momentum isobars involved will depend upon the spin of the external particles: the higher the external spins, the higher the number of different angular momentum states contributing to C_J. So C_J is a quadratic form in the coupling constants of nearby angular momentum isobars with the initial and final particles. The solution of the system (5.3) in terms of the coupling constants represents the general solution of the superconvergence equations in the model.

A formal solution can be found rather easily: one can express the partial waves as function of the amplitudes A_i, $\bar{A}_i$. The solution of the system (5.3) is obtained then putting in such expression all $\bar{A}_i = 0$. However such solution is not yet satisfactory.

We have further conditions, the positiveness of imaginary parts of scattering partial waves, or, in other words, the requirement that coupling constants be real.

Looking at some examples, we have found that the solutions one can obtain by taking simple polynomials for $\bar{a}_i(t)$ do not satisfy the positiveness conditions, and give rise to fake solutions of the superconvergence problem with imaginary coupling constants.

Let us investigate a model of π-N scattering in which internal symmetry variables are not present, and all Regge trajectories are lower than zero. We have

$$T = A + (\gamma Q)B$$

and the superconvergence sum rule holds

$$\int \mathrm{Im}\, B(s,t)ds = 0 \quad . \qquad (5.4)$$

Let us saturate with single mass intermediate states. Then

$$\mathrm{Im}\, A(s,t) = a(t)\, \delta(s-s_o) \quad ,$$

$$\mathrm{Im}\, B(s,t) = b(t)\, \delta(s-s_o) \quad ,$$

and (5.4) becomes

$$b(t) = 0 \quad . \qquad (5.5)$$

We may satisfy (5.5) with intermediate states of limited spin. A simplest saturation satisfying (5.5) is obtained by putting as intermediate states a $(\frac{1}{2})^+$ and $(\frac{1}{2})^-$ isobar.

The contribution of these states is easily written as

$$f_+^2(\gamma K+m) + f_-^2\gamma_5(\gamma K+m)\gamma_5 = f_+^2 + f_-^2(\gamma K-m)$$

where K is the total momentum, $K = \frac{P_1+P_2}{2} + \frac{Q}{2}$.
Eq. (5.5) then becomes

$$f_+^2 + f_-^2 = 0 \quad . \qquad (5.6)$$

Since f_+ and f_- are real, they must vanish. We see therefore that $(\frac{1}{2})^\pm$ particles formally saturate (5.4), but we get a fake solution.

An analogous situation shows up in a scattering process of a vector particle by a scalar one. Let us put the same scalar and vector particles (with equal masses and transforming as the same representation of a symmetry group) as intermediate states, introducing the simple couplings

(SSV):

$$g_S \; S^\alpha \; V_\mu^\beta \; \partial_\mu \; S^\gamma \; f_{\alpha\beta\gamma} \quad ,$$

(SVV):

$$g_V \; S^\alpha \; V_\mu^\beta \; V_\mu^\gamma \; f_{\alpha\beta\gamma} \quad ,$$

where α, β, γ are internal symmetry indices.

The contribution to the scattering $S^\alpha + V^\beta \rightarrow S^{\alpha'} + V^{\beta'}$ coming from the intermediate scalar takes the form

$$g_S^2(\varepsilon_1 K)(\varepsilon_2 K) f_{\alpha\beta\gamma} \; f_{\alpha'\beta'\gamma} \; \delta(s-s_o) \; ,$$

whereas the vector contribution is

$$- g_V^2[(\varepsilon_1\varepsilon_2) - \frac{(\varepsilon_1 K)(\varepsilon_2 K)}{s_o}] f_{\alpha\beta\gamma} f_{\alpha'\beta'\gamma} \delta(s-s_o) .$$

The complete direct and crossed contributions are thus given by

$$\{- g_V^2(\varepsilon_1\varepsilon_2) + (g_S^2 + \frac{g_V^2}{s_o})(\varepsilon_1 P)(\varepsilon_2 P)\}\{f_{\alpha\beta\gamma}f_{\alpha'\beta'\gamma}\delta(s-s_o) -$$

$$- f_{\alpha'\beta\gamma}f_{\alpha\beta\gamma'}\delta(u-s_o)\} .$$

In order to achieve superconvergence we must evidently have

$$g_S^2 + \frac{g_V^2}{s_o} = 0 \quad , \qquad (5.6')$$

and hence saturation with scalar and vector gives only the trivial result. Such examples lend to some amusing consideration.

In the latest example, we would have achieved superconvergence by introducing a simple "partiole" with propagator $g_{\mu\nu}/(s-s_o)$. Now, this propagator corresponds to a four vector field without imposition of auxiliary conditions: in the center of mass system the space components represent the spin one particle, and the time component represents the spin zero particle. However it is well known that the above interpretation of such a field is physically untenable, since the states would not possess a positive definite norm in general, and this is reflected in the fact that (5.6') cannot be satisfied with real non vanishing coupling constants. Analogously, (5.5) could be satisfied through the exchange of a spinor particle with propagator $1/(s-s_o)$, whose lower components in the center of mass system represent a $(\frac{1}{2})^-$ particle, but again with the same difficulties. The above observations are connected with the fact that four vector and four spinor are non unitary representations of the Lorentz group.

One can generalize the above argument rather easily, and show, at least in the case of π-N scattering, that no solution of the superconvergence requirements exists if the spin of the intermediate states is limited from above, the reason being in the positive definiteness of the imaginary parts of partial waves. The proof is easy in the equal mass case. Indeed, from the condition $\mathrm{Im}\, B(\nu,t) = 0$, using standard formulas, we get the condition on the partial wave amplitudes *

$$(f_{J,\pm} = f_{J=\ell\pm 1/2})$$

$$(E-m)\,\mathrm{Im}\, f_{J,+} + (E+m)\,\mathrm{Im}\, f_{J,-} = (E-m)\,\mathrm{Im}\, f_{J+1,-} + \\ + (E+m)\,\mathrm{Im}\, f_{J+1,+} \; . \qquad (5.7)$$

Suppose now $\mathrm{Im}\, f_{J,\pm} = 0$, $J > J_o$. Then

$$(E-m)\,\mathrm{Im}\, f_{J_o,+} + (E+m)\,\mathrm{Im}\, f_{J_o,-} = 0 \quad ,$$

and hence, because of the positive definiteness, $\mathrm{Im}\, f_{J_o,\pm} = 0$. Then, using (5.7) for $J < J_o$, for the same reasons of positiveness all phase shifts must vanish. We arrive at the conclusion that the only possible saturation of superconvergence through intermediate states of limited spin is the trivial one, with A=0 and B=0. This result does not depend upon the assumed equal mass isobaric model, and it can be shown to hold in general [4]. A similar general theorem can be shown to hold in π-ρ scattering [14].

So, from the previous discussion, we are led to draw the following conclusions:

1. Superconvergence equations (at least in the equal mass case) are saturated by means of the particles con-

---

* The reasons of Sakita and Wali argument referred in the previous section are clearly seen from (5.7).

tained in a single representation of the Lorentz group.

2. The usual non unitary finite dimensional representations lead us to unacceptable solutions where imaginary coupling constants are required.

3. The need for an infinite number of intermediate spins can be related to the fact that in order to saturate superconvergence with real coupling constants we need unitary representations of the Lorentz group, which contain particles with an infinite set of spins, and all states possess a positive norm.

Let us end up this section by giving a rough idea of how superconvergence + positiveness conditions can be fulfilled by a saturation through intermediate states of all spins and equal masses contained in a unitary (infinite dimensional of course) representation of the homogeneous Lorentz group [15].

To describe particles of higher spin one usually introduces a tensor field obeying certain auxiliary conditions, which insure the irreducibility of the representation of the little (rotation) group. For instance to describe a spin one particle, one introduces a four field Φ_μ , with the subsidiary condition

$$\frac{\partial \Phi_\mu}{\partial x_\mu} = 0 \quad . \tag{5.8}$$

(5.8) insures that in the rest system of the particle, no spin zero is present. On the other hand it is well known that it is not possible to interpret the spin zero part as a true particle, since the space of states would show an indefinite metric.

Of course one has complete freedom of complicacy in describing a particle of spin J, provided one introduces a suitable tensor of rank $\geq$J and suitable auxiliary conditions.

We may also describe a spin J particle through a unitary representation of the Lorentz group; indeed if a unitary representation contains the spin J, it is just a matter of projection to extract that spin from all objects contained therein.

We may define a "polarization" index α corresponding to a (J,m) couple contained in a unitary representation. Let us remember that a unitary representation contains all integer (or half integer) spins larger than a lowest value. Then the condition insuring us that we actually describe a spin J particle, is that in the rest system only the 2J+1 components belonging to the J polarization survive. Of course, the same representation can be used to describe independently particles of all spins contained therein . No dynamical assumption is implied, up to now we use a representation as a pure kinematical device which allows to describe contemporarily all spins which can be contained in it.

It is clear that such an algorithm is particularly wasteful when we are interested in phenomena involving particles of definite spin. However, such description is the most suitable candidate when we wish to treat simultaneously problems involving infinite particles with different spin.

From this kinematical point of view, we do not assign any fundamental role to the fact that infinite particles are contained in the same representation: a representation is just an array of states from which we can pick what we need. In principle, we do not require that particles which we pick up have equal masses. Thus, we characterize a particle by a certain momentum p and an index $\alpha = (j,m)$.

According to this formalism, we may write a general scattering amplitude as

$$T^{\alpha_1\alpha_2\alpha_3\alpha_4}(p_1,p_2,p_3,p_4) \qquad (5.9)$$

where α_i and p_i are the polarization and momentum of the i-th particle. To interpret (5.9) as a scattering amplitude of particles of spin and magnetic quantum numbers $j_1 m_1$, $j_2 m_2$, $j_3 m_3$, $j_4 m_4$ we define the function

$$F^{j_1 m_1, j_2 m_2, j_3 m_3, j_4 m_4}(p_1, p_2, p_3, p_4) =$$

$$= \sum_{\alpha_1 \alpha_2 \alpha_3 \alpha_4} u_{\alpha_1}^{j_1 m_1}(p_1) u_{\alpha_2}^{j_2 m_2}(p_2) u_{\alpha_3}^{j_3 m_3}(p_3) u_{\alpha_4}^{j_4 m_4}(p_4) \times$$

$$\times T^{\alpha_1 \alpha_2 \alpha_3 \alpha_4} . \qquad (5.10)$$

This formula is the generalization of a well known procedure; for instance in the scattering of spin 0 by spin 1/2 particles one defines a $T_{\alpha\beta}$ matrix, where α and β run from 1 to 4, and then contracts the indices as in (5.10)

$$\sum_{\alpha,\beta=1}^{4} u_\alpha^i(p)\ u_\beta^i(p')\ T_{\alpha\beta}$$

obtaining the matrix element between spinor states i and j with momenta p and p'. In (5.10) $u_\alpha^{jm}(p)$ is the $\alpha = (j'm')$ component of a "spinor" which, in its center of mass, possesses only the $\alpha = j, m$ components, and of course acquires all other components through the Lorentz transformation, and is an infinite dimensional analogoue of the Dirac spinors.

The spinors $u_\alpha^{jm}(p)$ have been investigated in the mathematical literature. It is not so difficult to construct them explicitly in a simple representation (Majorana) where they obey a Dirac-like equation

$$[\gamma p - m(j+ \tfrac{1}{2})]u^j = 0 \ , \ \text{and} (p^2-m^2)u^j=0 \quad ,$$

the γ being infinite dimensional generalizations of the

Dirac matrices, with no anticommutation closure.

The function (5.9) can be decomposed into invariants in the same way as for usual amplitudes except that in this case their number will be infinite, and indeed they must correspond to scattering processes of particles endowed with all spins contained in the representation.

Up to now, we have done nothing new, rather we have packed in an elegant way an infinite number of actual scattering processes. (5.9), (5.10) would be very bad objects if we had in mind to discuss π-π scattering and then go home. However, things which are complicated in one formalism may become simple in a different one, and different formalism may lead themselves to perform entirely different approximations.

We wish to show that in the framework of this formalism it is easy to obtain a model of intermediate state contributions, and couplings, which insure superconvergence requirements and together reality of coupling constants. Such a model is elementary in this formalism, but is not elementary at all if interpreted from the usual point of view. We consider the scattering of particles with any spin on a scalar target, and suppose that all particles, external or intermediate, will have equal masses:

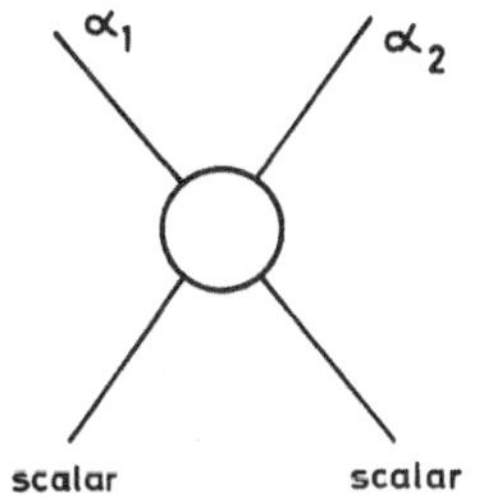

The imaginary part of the amplitude has the form

$$\mathrm{Im}\, T^{\alpha_1 \alpha_2} = t^{\alpha_1 \alpha_2}\, \delta(s-m^2) \quad . \qquad (5.11)$$

In this case superconvergence is achieved by putting to zero all amplitudes having a superconvergent asymptotic behaviour. In particular superconvergence is certainly insured if $t^{\alpha_1\alpha_2}$ has the form

$$t^{\alpha_1\alpha_2} = t\delta^{\alpha_1\alpha_2} \tag{5.12}$$

where the polarization indices of the particles with spin are contracted together and therefore the asymptotic behaviour of this amplitude has to be a Regge one, s^{α} , without further convergence factors.

Such a form for $t^{\alpha_1\alpha_2}$ can be easily obtained writing for the (α,β,scalar) vertex the coupling

$$g\,\delta_{\alpha\beta}\;. \tag{5.13}$$

Indeed, computing the imaginary part of the physical amplitude, obtained taking (5.12) between external spinors, due to exchange of all possible intermediate particles belonging to the representation, we have

$$u_{\alpha_1}^{j_1m_1} t^{\alpha_1\alpha_2} u_{\alpha_2}^{j_2m_2} =$$

$$= g^2 \sum_{j,m,\beta,\beta'} u^{j_1m_1}\delta_{\alpha_1\beta} u_{\beta}^{jm}\; u_{\beta'}^{jm}\delta_{\beta'\alpha_2} u_{\alpha_2}^{j_2m_2}\;. \tag{5.14}$$

Now, one can use the completeness relation

$$\sum_{j,m} u_{\beta}^{jm}(p)\; u_{\beta'}^{jm}(p) = \delta_{\beta\beta'} \tag{5.15}$$

where the two spinors are taken at the same momentum p, and then (5.14) becomes

$$t^{\alpha_1\alpha_2} = g^2\,\delta_{\alpha_1\alpha_2}$$

We have therefore seen that superconvergence is fulfilled by introducing all particles in the intermediate state belonging to the unitary representation, with the coupling (5.12). At the same time all couplings between physical particles are real, and here is the fundamental difference from the models of finite representations previously discussed. Indeed the coupling between two physical particles is given by

$$g\, u_\alpha^{j_1 m_1}(p_1)\, u_\alpha^{j_2 m_2}(p_2) \tag{5.16}$$

and such coupling is real for real g, because the representation is unitary.

The very simple T matrix (5.12) and coupling (5.16) are immediately realized to be complicated objects from the conventional point of view, and involve definite relations between vertices and scatterings of particles of different spins.

Finally we notice that a still large freedom is allowed. We did not write explicitly the dependence upon the particular unitary representation chosen, and indeed we may take any continuous superposition of unitary representations. Maybe this freedom has to do with the "fixed" t channel, and it is needed to reproduce the mass singularity spectrum in t. *

Let us finally observe that we are still far away from an approach to real physics. We are bound (at least for the moment) to consider equal mass case only, otherwise we cannot use completeness in (5.14). Still, even though a long road must still be covered before getting near to physics, we think that use of unitary representations of the Lorentz group will be the natural clue for the solution of superconvergence and, as we

* It is not difficult to see that form factors as (5.16) show the perturbative type threshold at $t=4m^2$, so that to reproduce a pole at $t=m^2$ a superposition of representations must be used.

will see, current algebra equations.

6. Current Algebra

In this section we shall discuss some features of current algebra sum rules, in particular the asymptotic behaviour of the amplitudes, and the comparison with the strong interaction sum rules [4]. Let us start by considering the following amplitude

$$T^{\alpha\beta}_{\mu\nu} = i\int d^4x\, e^{-iq_1x} \langle p_2|T(j^\alpha_\mu(x), j^\beta_\nu(o))|p_1\rangle$$

where $j^{\alpha,\beta}_\mu$ are currents, α and β internal symmetry indices, $|p_1\rangle$ and $|p_2\rangle$ scalar or pseudoscalar particle states (let us say pions).

We do not write explicitly possible Schwinger terms. On the other hand they would affect only non superconvergent amplitudes, as we shall see.

Kinematics is shown in fig. 6.1

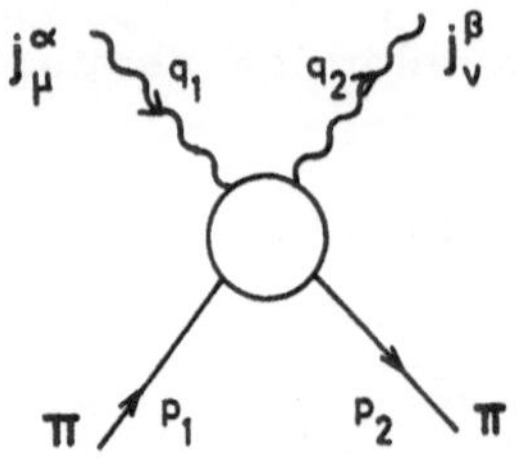

$$\begin{aligned} K &= p_1 - p_2 \\ P &= p_1 + p_2 \\ Q &= q_1 + q_2 \\ q_i^2 &= u_i \\ p_i^2 &= m_i^2 \\ t &= (p_1 - p_2)^2 \\ \nu &= \tfrac{1}{2}(P\,Q) \end{aligned}$$

Fig. 6.1

The relations with the definitions used in sect. 3 are

$q_1 \to k_1,\ q_2 \to -k_1',\ p_1 \to -k_2,\ p_2 \to k_2'$.

According to the technique developed in sect. 3 we can write $T_{\mu\nu}$ in the form

$$T_{\mu\nu} = A_1 P_\mu P_\nu + A_2 P_\mu q_{1\nu} + A_3 P_\mu q_{2\nu} + A_4 P_\nu q_{1\mu} + $$
$$+ A_5 P_\nu q_{2\mu} + A_6 q_{1\mu} q_{1\nu} + A_7 q_{1\mu} q_{2\nu} + A_8 q_{1\nu} q_{2\mu} + $$
$$+ A_9 q_{2\mu} q_{2\nu} + A_{10} g_{\mu\nu} ;$$
$$A_i = A_i(\nu, t, u_1, u_2) , \qquad (6.2)$$

where the invariant functions $A_1 \ldots A_{10}$ are devoid of kinematical singularities, and the above decomposition is also suggested by perturbation theory.

We are interested in the asymptotic behaviour of the A_i functions for large ν at fixed t. Now, a very important feature of this case is the appearance of terms giving for large ν an asymptotic behaviour in competition with the Regge one. Indeed we show that a hypothesis of pure Regge pole behaviour for all weak amplitudes is incompatible with the consequences of current algebra.

Let us suppose for an instant that a Regge pole behaviour could be accepted both for real and imaginary parts of the amplitudes A_i, i.e. let us suppose the behaviour for large ν

$$A_1 \sim \nu^{\alpha-2}$$
$$A_i \sim \nu^{\alpha-1} \qquad i = 2, 3, 4, 5,$$
$$A_i \sim \nu^{\alpha} \qquad i = 6, 7, 8, 9, 10. \qquad (6.3)$$

Using a standard procedure, we can calculate the quantity $q_{1\mu}\, T_{\mu\nu}$ from expansion (6.2):

$$q_{1\mu}\, T_{\mu\nu} = [\nu A_1 + u_1 A_4 + t_1 A_5] P_\nu +$$

$$+ \left[\nu A_2 + u_1 A_6 + t_1 A_8 + A_{10}\right] q_{1\nu} + \left[\nu A_3 + u_1 A_7 + t_1 A_9\right] q_{2\nu} \tag{6.4}$$

where $t_1 = \frac{1}{2}(u_1 + u_2 - t)$. According with the hypothesis (6.3), the coefficient of P_ν in (6.4) behaves as $\nu^{\alpha-1}$, those of $q_{1\mu}$, $q_{2\nu}$ as ν^α.

Now we may calculate $q_{1\mu}\ T_{\mu\nu}$ directly from (6.1), obtaining

$$q_{1\mu}\ T_{\mu\nu} = U_\nu + R_\nu \quad . \tag{6.5}$$

U_ν and R_ν are given by

$$U_\nu = i\int d^4x\ e^{-iq_1 x} \langle p_2 \left[T(D^\alpha(x), j^\beta_\nu(o))\right] p_1 \rangle \ ,$$

$$R_\nu = i\int d^4x\ e^{-iq_1 x} \langle p_2 | \left[j^\alpha_o(x), j^\beta_\nu(o)\right] | p_1 \rangle \delta(x_o) \ ;$$

$$D^\alpha = \partial_\mu\ j^\alpha_\mu \quad .$$

Now, U_ν is again an amplitude, where a vector current has been replaced by a divergence. We may expand U_ν into invariants:

$$U_\nu = B\ P_\nu + C_1\ q_{1\nu} + C_2\ q_{2\nu} \quad .$$

Again making a pure Regge hypothesis,

$$B \sim \nu^{\alpha-1} \quad , \quad C_{1,2} \sim \nu^\alpha \quad .$$

Let us investigate R_μ. The simplest form required by causality is

$$\delta(x_o)\left[j^\alpha_o(x), j^\beta_\nu(o)\right] = J^{\alpha\beta}_\nu(x)\delta^4(x) \quad . \tag{6.6}$$

In particular, from current algebra hypothesis,

$$J^{\alpha\beta}_\nu(x) = i\ f^{\alpha\beta\gamma}\ j^\gamma_\nu(x) \quad .$$

As a consequence, the functional dependence of R_ν is $R_\nu = R_\nu(p_1,p_2)$ and therefore, when decomposed into invariant functions, these cannot depend upon q_1^2,q_2^2.

$$R_\nu(p_1,p_2) = P_\nu\ \beta(t) + (q_1 - q_2)_\nu\ \gamma(t) \quad . \qquad (6.7)$$

Possible δ-gradients in (6.4) would introduce some separate dependence on q_1,q_2. From (6.5), equating the coefficients of P_ν, we get

$$\nu A_1 + u\, A_{1\,4} + t\, A_{1\,5} = B + \beta(t) \quad . \qquad (6.8)$$

Let us examine (6.8) for the imaginary parts. Then no equal time contribution appears (Im $\beta(t) = 0$) and (6.8) for the imaginary parts is perfectly consistent with a pure Regge behaviour. (6.8) is however contradictory with Regge behaviour for the real parts. Indeed if we take a channel with $\alpha(o) < 1$ (as it is for T=1 in the crossed t channel) then the constant term for large ν, $\beta(t)$, has no partner. We must drop the hypothesis of pure Regge behaviour for real and imaginary parts of amplitudes involving weak currents. The origin of asymptotic non Regge terms can be traced back to the mixed nature of the problem (weak hadron currents are considered), which is of second order in weak interaction and exact in strong interactions. In other words if we switch off all strong interaction effects, the amplitudes describing scattering processes of hadrons do of course vanish, whereas the amplitudes appearing in the decomposition (6.2) reduce to their Born approximation (in weak interactions).

Therefore, when we investigate the damping of strong interaction effects at high energies according to the Regge pole theory, we must expect that some terms, related to the structure of weak currents, may survive. This will be particularly important for the superconvergent amplitudes $\sim \nu^{\alpha-2}$, where, because of the strongest damping of the Regge pole contribution, we can expect the weak interaction effect to be dominant. Indeed the Born graphs introduce in all amplitudes A_i a term behaving asymptotically as ν^{-1}, which is the dominant one for superconvergent amplitudes (remember $\alpha < 1$) .

On the other hand, if strong interaction effects are switched off, only the real part will survive, because the absorptive part is directly related to multiple production, whereas the remaining Born term of weak interaction is essentially real.

The previous discussion leads us to the following hypothesis: whereas strong interaction amplitudes will be assumed to have a Regge pole behaviour, for amplitudes involving combinations of both particles and currents the asymptotic Regge behaviour will be assumed only for absorptive parts.

Of course the knowledge of the high energy behaviour of the absorptive parts, together with an appropriate hypothesis about subtractions, leads to predictions about the real part of the amplitude.

We shall assume that the invariants A_i of eq. (6.2), which are suggested both by the analysis of sect. 3 and by perturbative considerations, require no more subtractions than those needed to ensure convergence of the dispersion integrals.

We shall use here again a theorem, whose proof is given in App. 1:

Let us consider an analytic amplitude $F(\nu)$, with the singularities required by unitarity, and whose imaginary part $\mathrm{Im}\, F \sim \nu^b$ as $\nu \to \infty$. For this amplitude we assume only the number of subtractions required by convergence. Then

a) for $b > -1$ also $\mathrm{Re}\, F \sim \nu^b$ (non superconvergent case);

b) for $b < -1$ we can write (superconvergent case)

$$\mathrm{Re}\, F = \sum_1^n{}_p \frac{a_p}{\nu^p} + \Phi(\nu) ,$$

$$\Phi(\nu) \sim \nu^b , \qquad b = -n - \varepsilon,\ 0<\varepsilon<1 ,$$

$$n = \text{non negative integer} , \qquad (6.9)$$

and

$$a_p = -\frac{1}{\pi} \int \nu^{p-1} \text{ Im } F(\nu) d\nu .$$

In the case of strong interactions, where we introduce the additional assumption of Regge behaviour for the real part, the previous theorem shows that this requirement is automatically satisfied for $b > -1$, but it is equivalent to the requirement of the superconvergence identities

$$a_1 = a_2 = \ldots = a_n = 0$$

for $b < -1$.

Let us now go back to the form for $T_{\mu\nu}$ given by eq. (6.2). In order to discuss the asymptotic behaviour of the specific amplitudes we need to consider explicitly the dependence of $T_{\mu\nu}$ on the internal variables.

Let us discuss the isospin case where we define the amplitudes $A_i^{(o)}$, $A_i^{(1)}$, $A_i^{(2)}$ for $T = 0, 1, 2$ in the t channel (generalization to SU(3) is straightforward). We discuss the three cases separately.

For $T = 0$, no $A_i^{(o)}$ amplitude is really superconvergent.

The most interesting case corresponds to isospin one, when only the first amplitude $A_1^{(1)}$ is superconvergent. According to eq. (6.9) the real parts of the amplitudes will have the following asymptotic behaviour:

$$A_1^{(1)} \sim \text{const.}\nu^{\alpha-2} + \frac{a_1}{\nu} ,$$

$$A_i^{(1)} \sim \text{const.}\nu^{\alpha-1} , \qquad i = 2, 3, 4, 5,$$

$$A_i^{(1)} \sim \text{const.}\nu^{\alpha} \quad , \quad i = 6, 7, 8, 9, 10 \ . \qquad (6.10)$$

Eqs. (6.10) reflect the overall philosophy of this section. All amplitudes A_i may have non Regge contributions of order $1/\nu$. However these contributions shall dominate only in the superconvergent amplitude $A_1^{(1)}$. This can be understood by the physical observation that if we switch off strong interactions, we are left with the Born approximation in weak interactions, which, for the perturbative invariants A_i, consists only of poles.

We are now faced with the problem of identifying the constant a_1 .

This can most easily be performed through (6.8) and we get

$$\lim_{\nu\to\infty} \nu A_1^{(1)} = \beta(t) \qquad (6.11)$$

leading to the Fubini, Dashen - Gell-Mann sum rule

$$\int \text{Im } A_1^{(1)}(\nu,t,u_1,u_2)\, d\nu = -\ \beta(t) \qquad (6.12)$$

Before discussing in detail (6.12), let us finally consider the case of T = 2. In this case, assuming $-1 < \alpha^{(2)} < 0$, the amplitudes $\nu A_1, A_2, A_3, A_4, A_5$ have a superconvergent Regge behaviour, and the correct crossing properties to yield sum rules. To identify their asymptotic part, we should write a set of equations analogous to (6.8), but it is easy to find out that, because of current algebra structure, no term from an equal time commutator can give a $\frac{1}{\nu}$ behaviour, so that these amplitudes satisfy a strong interaction type sum rule:

$$\int \text{Im } A_i^{(2)}(\nu,t,u_1,u_2)d\nu = 0 \quad .$$

Going back to (6.12), many questions can be posed. How can the l.h.s. reproduce the analytic properties of

$\beta(t)$, which is after all a form factor? Indeed, the only state contributing to $\beta(t)$ has $J = 1$ in the t channel, while $A_1(t)$ gets contributions from states with $J \geq 2$. Second question, how it comes that the r.h.s. of (6.12) is independent of u_1 and u_2 ? In discussing these ideas we follow the paper by Bronzan, Gerstein, Lee and Low [16] (for the fixed pole at J = 1 see also V. Singh [17]. We may define a partial wave expansion of $A_1(\nu,t)$ in the t channel:

$$A_1(\nu,t) =$$

$$= \sum_{J=2}^{\infty} (2J + 1)F_J(t,u_1,u_2)P''_J(\cos\theta_t) \quad . \tag{6.13}$$

Performing the Watson-Sommerfeld transform and supposing a moving pole to be dominant, we would have for the asymptotic behaviour at large ν

$$A_1 \sim \frac{\gamma\ (t)}{\sin\pi\alpha(t)}\ (2\alpha+1)\ \alpha(\alpha-1)\ \frac{\Gamma(\alpha+1/2)}{\Gamma(\alpha+1)}\ (-2\cos\theta)^{\alpha-2}$$

Now, if we continue analytically (6.12) from the region of spacelike t to time-like t, the pole in $\beta(t)$ at $t = m_\rho^2$ has no way of being reproduced by the l.h.s. of (6.12). Indeed the asymptotic part of the integrand, which would cause the integral to diverge for $\alpha \to 1$(i.e. $t \to m_\rho^2$) has a vanishing coefficient. A moving pole mechanism is therefore not compatible with (6.12) continued up to $t = m_\rho^2$.

For (6.12) to be valid, it is necessary to have a fixed pole at J = 1 for the t channel partial amplitude F_J. Of course such pole is not physically seen since the expansion in (6.13) begins at J = 2.

If we put

$$F_J(t,u_1,u_2) = \frac{\Gamma_J(t,u_1,u_2)}{(J-1)(J-\alpha)} \quad . \tag{6.14}$$

where Γ_J is analytic and non vanishing in the neighbourhood of $J = 1$ and of $J = \alpha(t)$, then the asymptotic behaviour of A_1, by Watson-Sommerfeld transform, is given by the fixed pole contribution so long as $\alpha(t) < 1$, i.e. $t < m_\rho^2$.

Such a fixed pole contributes only to the real part of the amplitude, and the imaginary part presents a Regge asymptotic behaviour without the coefficient $(\alpha-1)$. So the mechanism of production of singularity of $\beta(t)$ at $t = m_\rho^2$ through divergence of the integral at l.h.s. of (6.12), suggested by Fubini and Segré [6], works indeed. As $\alpha \longrightarrow 1$ (6.11) loses sense and the analytic continuation of the integral for $t > m_\rho^2$ is given by $\beta(t)$. (6.11) expresses the requirement that the residue of the fixed pole in the J plane is $\beta(t)$. The preceding discussion sketches the correct translation in angular momentum language of our preceding dispersive discussion about asymptotism for weak amplitudes, and the presence of a fixed J pole is the reminiscence of the mixed nature of our problem, where weak interactions are treated in the lowest order.

The mechanism of the fixed pole is supported by an explicit calculation by Bronzan, Gerstein, Lee and Low, who show that the sum rule (6.12) holds in the ladder approximation of vector current-pion scattering; in this model the presence of a fixed pole at $J = 1$ is exhibited clearly.

In order to discuss the second problem, namely how the r.h.s. of (6.12) is independent of the current "masses" u_1 and u_2, we resort again to the BGLL-paper.

Let us deal with conserved current, i.e. $D^\alpha = 0$, and let us take the particular kinematical configuration $u_1 = 0$, $u_2 = t$.

Then (6.8) for the imaginary part becomes

$$\nu \mathrm{Im}\, A_1(\nu,t,0,t) = 0,$$

$$\text{Im } A_1(\nu,t,0,t) = C(t)\ \delta(\nu) \quad .$$

The intermediate particle which contributes with the above form is the pion, the same as the initial one, and we may calculate its contribution exactly through the diagram

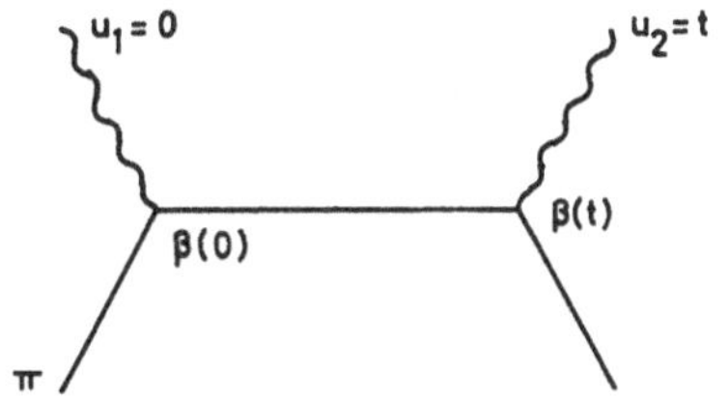

One then obtains

$$C(t) = \beta(t)\ \beta(o) = \beta(t)$$

under the hypothesis of conserved current, and (6.12) is established. Let us notice that such particular kinematical configuration corresponds to the consideration of a charge-current commutator, for which the presence of additional Schwinger terms is extremely unlikely.

Now let us continue from $A_1(\nu,t,0,t) \equiv A_1(\nu;o)$ to $A_1(\nu,t,u_1,t) \equiv A_1(\nu;u_1)$. If we suppose that, as function of u_1, A has poles and cuts corresponding to possible intermediate hadron states, we may write

$$A_1(\nu;u_1) - A_1(\nu;o) = u_1 \frac{\rho_o(\nu)}{u_1 - u_o} + u_1 \int du' \frac{\rho(\nu,u')}{u'-u_1} \quad ,$$

where the pole term symbolizes the contribution of a sharp mass u_o hadron state (there can be many) and the continuum term corresponds to many hadron states. $\rho_o(\nu)$ is proportional to an amplitude for creation of a hadron by a current on a pion, and this amplitude is of superconvergent type. Now, such an amplitude tends to zero if we switch off strong interactions, and therefore a pure Regge behaviour must be allowed for both real and imaginary parts of the amplitudes. Therefore ρ_o satisfies a strong interaction type superconvergence sum rule,

$$\int \mathrm{Im}\ \rho_o(\nu)d\nu = 0 \ .$$

The same reasoning holds for $\rho(\nu,u)$. Therefore

$$\int \mathrm{Im}\ A(\nu;u_1)d\nu =$$

$$= \int \mathrm{Im}\ A(\nu;o)d\nu + \frac{u_1}{u_1-u_o}\int \mathrm{Im}\ \rho_o(\nu)d\nu +$$

$$+ u_1\int \frac{du'}{u'-u_1}\int \mathrm{Im}\rho(\nu,u')d\nu = \int \mathrm{Im}\ A(\nu;o)d\nu = \beta(t)$$

(6.12) is then shown to hold in general, and the mechanism by which the r.h.s. is independent of u_i is clearly displayed. To write (6.12) in all its generality we just need charge-current commutator algebra, and the rest is superconvergence, having allowed for inelastic superconvergence.

We want finally to remark that a solution of current algebra conditions can be given by saturating with equal mass particles contained in a unitary representation of the Lorentz group, as in the case of strong interaction sum rules. Indeed the coupling of current with particles can be chosen to be

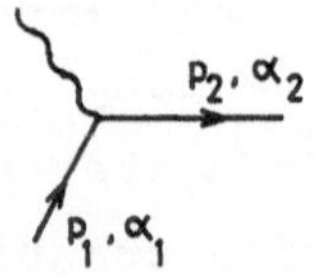

$$j^{(i)}_{\mu,\alpha_1,\alpha_2} = (p_1+p_2)_\mu\ \lambda_i \delta_{\alpha_1\alpha_2} ,$$

where λ_i an appropriate SU(2) or SU(3) matrix. Then again the completeness relation insures that equal time commutation relations between charge densities are satiesfied. This solution corresponds to that obtained by R. Dashen and M. Gell-Mann starting from a different point of view, [18].

The form factors are given by

$$\Sigma_\alpha\, u_\alpha^{J_1 m_1}(p_1)\, u_\alpha^{J_2 m_2}(p_2)$$

and satisfy the spectral representation

$$\beta(t) = \frac{1}{\pi} \int_{4m^2}^{\infty} \frac{\rho(t')}{t'-t}\, dt'$$

with the correct perturbative threshold.

Appendix 1

In this appendix we show that, as simple consequence of a fundamental property of the Stieltjes-Hilbert transform, one can deduce the asymptotic behaviour of the real part of a scattering amplitude once the behaviour of its imaginary part is known, provided the usual analytic properties are assumed.

First of all we remember that an Abelian theorem states that being

$$F(\nu) = \frac{1}{\pi}\, P \int_{-\infty}^{+\infty} \frac{f(\nu')}{\nu'-\nu}\, d\nu' \quad , \qquad (1)$$

if $f(\nu)$ behaves asymptotically for $\nu \to \infty$ as const.ν^α with $-1<\mathrm{Re}\,\alpha<0$, then the transformed function $F(\nu)$ enjoys the same asymptotic behaviour for $\nu \to \infty$.

As a simple consequence we can now state the following theorem: given an analytic regular function satisfying the asymptotic requirement

$$f(\nu) \sim C\nu^{\alpha} \quad ,$$

$\mathrm{Re}\alpha = -n-\varepsilon$, n non negative integer, $0<\varepsilon<1$, then the asymptotic expansion for large ν of the Stieltjes-Hilbert transform is given by

$$F(\nu) = \sum_{1}^{n} {}_p \frac{a_p}{\nu^p} + \mathrm{const.}\,\nu^{\alpha} \quad , \tag{2}$$

where

$$a_p = -\frac{1}{\pi} \int_{-\infty}^{+\infty} \nu^{p-1} \cdot f(\nu)d\nu \quad . \tag{3}$$

The proof goes through the use of the elementary identity

$$\frac{1}{\nu'-\nu} = -\frac{1}{\nu} \left[\sum_{o}^{n-1} {}_p \left(\frac{\nu'}{\nu}\right)^p + \frac{(\nu'/\nu)^n}{1-\nu'/\nu} \right] . \tag{4}$$

Indeed from (4) it follows

$$F(\nu) = \frac{1}{\pi} P \int_{-\infty}^{+\infty} \frac{f(\nu')}{\nu'-\nu} d\nu' = -\frac{1}{\pi} \sum_{1}^{n} {}_p \frac{1}{\nu^p} \int_{-\infty}^{+\infty} \nu'^{p-1} f(\nu')d\nu' +$$

$$+ \frac{1}{\nu^n} \frac{1}{\pi} P \int_{-\infty}^{+\infty} \frac{\nu'^n f(\nu')}{\nu'-\nu} d\nu' \quad . \tag{5}$$

Under the stated hypothesis on $f(\nu)$, all the integrals are convergent; besides, the transforming function in the last integral is

$$\nu^n f(\nu) \underset{\nu\to\infty}{\sim} \nu^{-\varepsilon} \quad ,$$

so that its transform has the same behaviour and the theorem is proved.

In conclusion, given an amplitude $A(\nu,t)$ for which

$$\mathrm{Re}\, A(\nu,t) = \frac{1}{\pi} P \int_{-\infty}^{+\infty} \frac{\mathrm{Im}\, A(\nu',t)}{\nu'-\nu} d\nu'$$

holds (implying of course $\mathrm{Im}\, A(\nu,t) \sim C(t)\nu^{\alpha(t)}$, Re $\alpha(t) < 0$, then the asymptotic behaviour of Re A (ν,t) is obtained by the above theorem. Moreover, if

$$\mathrm{Im}\, A(\nu,t) \sim C(t)\nu^{\alpha(t)} , \quad 0 < \mathrm{Re}\, \alpha(t) < 1 ,$$

writing a dispersion relation once subtracted,

$$\mathrm{Re}\, A(\nu,t) = \mathrm{Re}\, A(\nu_o,t) + \\ + (\nu-\nu_o) \frac{1}{\pi} P \int_{-\infty}^{+\infty} \frac{\mathrm{Im}\, A(\nu' t)}{(\nu'-\nu_o)(\nu'-\nu)} d\nu'$$

we recognize immediately that also

$$\mathrm{Re}\, A(\nu,t) \underset{\nu\to\infty}{\sim} C'(t)\nu^{\alpha(t)}$$

since the transforming function, $\mathrm{Im}\, A/(\nu-\nu_o) \sim \nu^{\alpha-1}$ and then the same asymptotic behaviour holds for its transform.

References

1. V. de Alfaro, S. Fubini, G. Furlan and C. Rossetti, Phys. Lett. 21, 576 (1966). The idea that strong interaction sum rules can be obtained from asymptotic limits has been independently developed by L. D. Soloviev, Sov. Journ. of Nucl. Phys. 3, 131(1966).
2. K. Hepp, Helv. Phys. Acta, 37, 55 (1964). See, also for general references, the paper by G. C. Fox, "Methods for constructing invariant amplitudes free from kinematical singularities and zeros". Cavendish Lab. preprint, Nov. 1966.
3. T. L. Trueman, Phys. Rev. Lett. 17, 1198 (1966). See also R. Odorico, University of Trieste preprint.
4. V. de Alfaro, S. Fubini, G. Furlan and C. Rossetti,

"Superconvergence and current algebra", University of Torino preprint, Dec. 1966.

5. M. Gell-Mann, S. Frautschi and F. Zachariasen, Phys. Rev. 126, 2204 (1962).
6. S. Fubini and G. Segré, N. Cim. 45, 641 (1966).
7. M. Gell-Mann, D. Sharp and W. G. Wagner, Phys. Rev. Lett. 8, 261 (1962).
8. P. H. Frampton and J. C. Taylor, "Superconvergence sum rules in $\rho\pi$ scattering" Oxford preprint.
9. P. H. Frampton, "The chirality commutator and vector mesons", Oxford preprint.
10. G. Altarelli, F. Buccella and R. Gatto, Phys. Lett. 24, 57 (1967); P. Babu, F. J. Gilman and M. Suzuki, Phys. Lett. 24, 65 (1967); B. Sakita and K. C. Wali, Phys. Rev. Lett. 18, 31 (1967).
11. The problem of connection with group results is the object of interesting investigations; see K. Bardakci and G. Segré, "The algebraic structure of superconvergence relations", preprint; R. Oehme, Phys. Lett. 21, 567 (1966) and 22, 207 (1966); R. Oehme and G. Venturi, University of Chicago preprints EFINS 66-84 and 67-4.
12. R. D'Auria and V. de Alfaro, University of Torino preprint, Dec. 1966.
13. L. D. Soloviev, Sov. Journ. Nucl. Phys. 3, 188 (1966); I. G. Aznaurian and L. D. Soloviev, Sov. Journ. of Nucl. Phys. 4, 615 (1966); V. A. Matveev, B. N. Struminski and A. N. Tavkhelidze, Phys. Lett. 23, 146 (1966); I. Kahn, Superconvergence relations for pion photoproduction and pion scattering, University of Edinburgh preprint; H. Harari, Phys. Rev. Lett. 18, 319 (1967); S. N. Biswas, J. Dhar and R. P. Saxena, "Coupling constant sum rules from fixed momentum transfer dispersion relations in $\pi+N \rightarrow \rho+N$ ", University of Delhi preprint; H. F. Jones and M. D. Scadron, "U(6,6) and superconvergent sum rules for $\pi+N \rightarrow \pi+N^*$ scattering",

Imperial College preprint; R. J. Rivers, "The meson spectrum and superconvergence sum rules for the 2^{++} mesons", Imperial College preprint.

14. C. Rebbi, private communication .
15. S. Fubini, talk at the IV. Coral Gables Conference, January 1967.
16. J. B. Bronzan, I. S. Gerstein, B. W. Lee and F. E. Low, Current algebra and non-Regge behaviour of weak amplitudes II, preprint.
17. V. Singh, Phys. Rev. Lett. 18, 36 (1967).
18. M. Gell-Mann, lectures at the International School of Physics, Erice, 1966, Cal. Tech. preprint.

DIVERGENCE EQUATIONS AND CURRENT ALGEBRA†

By

M. NAUENBERG

Division of Natural Sciences
University of California
Santa Cruz,USA

I. Introduction. Review of Current Algebra

The main topic which I plan to cover in these lectures is the contributions of electromagnetic and weak interactions to the divergence of the hadron currents. The corresponding divergence equations lead to an alternative approach to the well known method of current algebra and can be shown to imply equal-time current-current commutators. You may wonder then why we should consider this topic, in particular since the equivalence to the conventional approach can be demonstrated. One reason is that we are trying to extend these ideas to include dynamics of the hadrons and it is not a priori clear which view point leads more readily to useful generalization. Another reason is that the divergence method sheds light on certain confusion which has arisen in connection with the singular behaviour of equal time commutators of currents, the so-called

† Lect re given at the VI. Internationalen Universitätswochen f.Kernphysik,Schladming,26 February-11 March 1967.

Schwinger term.

Finally the divergence conditions help clarify the underlying assumptions of current algebra.

Let me begin with a brief review of current algebra. As you know, the internal symmetries of the strong interactions of hadrons together with Lorentz invariance suggest the existence of local vector currents $j_\mu^{V,\alpha}(x)$ such that

$$J_\alpha^V = \int d^3x \; j_o^{V,\alpha}(\vec{x},t) \tag{1.1}$$

are the generators (sometimes called charges) of the symmetry group. These generators satisfy commutation relations

$$[J_\alpha^V, J_\beta^V] = i \, f_{\alpha\beta\gamma} \, J_\gamma^V \tag{1.2}$$

where $f_{\alpha\beta\gamma}$ are the structure constants which characterize the group.

In the case of an exact symmetry, say SU(2) in the absence of electromagnetic and weak interaction forces, the hamiltonian H and the S matrix of the hadrons are invariants and therefore commute with the generators of this group. Furthermore, the generators are then time independent:

$$\frac{dJ_\alpha^V}{dt} = i\,[H, J_\alpha^V] = 0 \tag{1.3}$$

According to eq. (1.1)

$$\frac{dJ_\alpha^V}{dt} = \int d^3x \; \frac{\partial j_o^{V,\alpha}(\vec{x},t)}{\partial t} = \int d^3x \; \partial^\mu j_\mu^{V,\alpha}(\vec{x},t) \tag{1.4}$$

so that eq. (1.3) is satisfied if the four divergence of the currents j_μ^α vanish,

$$\partial^\mu j_\mu^{V,\alpha}(\vec{x},t) = 0 \tag{1.5}$$

It is well known that for SU(2) many relations among transition amplitudes for the hadrons have been deduced from purely group theoretical considerations irrespective of the dynamics. Actually electromagnetic and weak interactions are not SU(2) symmetric, but because of their relative weakness, these interactions can usually be neglected. However, if we want to evaluate specific effects due to symmetry breaking interactions (for example the electromagnetic mass splittings of hadrons) we must extend our previous considerations.

If part of the hamiltonian does not commute with a particular generator J_α^V, then it is time dependent. Electromagnetic interactions for example do not commute with the generators J_α^V of SU(2) for $\alpha = 1,2$; consequently $\frac{d}{dt} J_{1,2}^V$ must be proportional to the electromagnetic charge. Another well-known example is the approximate SU(3) symmetry of the strong interaction. In this case it is not even clear that we can isolate a part of the strong interaction which is SU(3) symmetric. However, we can assume that the commutation relations among the time dependent generators $J_\alpha(t)$, eq. (1.2), taken at equal times, remain valid.

In order to include weak interaction effects we must consider also axial vector currents $j_\mu^{A,\alpha}$ for the hadrons, which are assumed to transform under SU(3) like the vector currents. Hence, these axial currents satisfy commutation relations

$$[J_\alpha^V(t),\ j_\mu^{A,\beta}(\vec{x},t)] = i\, f_{\alpha\beta\gamma}\ j_\mu^{A,\gamma}(\vec{x},t) \quad . \tag{1.6}$$

In analogy with the case of vector currents we define also an axial charge

$$J_\alpha^A(t) = \int d^3x\ j_o^{A,\alpha}(\vec{x},t) \quad . \tag{1.7}$$

In this case, however, J_α^A is time dependent even in the absence of electromagnetic and weak interactions, and of SU(3) symmetry breaking forces. It is assumed [1] that

$$\partial^\mu j_\mu^{A,\alpha} = C\phi^\alpha \tag{1.8}$$

where C is a constant, and ϕ^α is the pseudoscalar meson field. This is known as the hypothesis of the partially conserved axial current PCAC and leads to the famous Goldberger-Treiman relation. As we shall see later on, the real content of eq. (1.8) is the fact that the divergence of the axial current is a good interpolating field for the pseudoscalar mesons.

The next interesting question to consider is the equal time commutator between axial charges J_α^A. Gell-Mann's fundamental assumption [2] states that this commutator is proportional to the vector charges J_α^V,

$$[J_\alpha^A(t),\ J_\beta^A(t)] = i\ f_{\alpha\beta\gamma}\ J_\gamma^V(t) \tag{1.9}$$

In this sense the axial and vector charges form a closed algebra. Note that the commutation relation between axial charges, eq. (1.9), is a very special assumption which is not generally satisfied in field theory models of the hadrons, while the other commutators, eq. (1.2) and (1.6), are always valid. An example which does satisfy eq. (1.9) is the quark model [2] in which we assume a fundamental spinor field $\psi(x)$ which transforms like the basic representation of SU(3). The vector and axial currents are then defined in terms of $\psi(x)$,

$$j_\mu^{V,\alpha}(x) = \bar\psi(x)\gamma_\mu\ \lambda_\alpha\ \psi(x) \tag{1.10}$$

$$j_\mu^{A,\alpha}(x) = \bar\psi(x)\gamma_\mu\ i\ \gamma_5\ \lambda_\alpha\ \psi(x) \tag{1.11}$$

where λ_α are the 3 × 3 matrix generators of SU(3). Equal time commutators between two currents are obtained from the canonical commutation relations satisfied by ψ. We consider here only the commutators when one of the currents has a time component, and obtain,

$$[j_o^{V,\alpha}(x), j_\mu^{V,\beta}(y)]_{x_o=y_o} = i\, f_{\alpha\beta\gamma} j_\mu^{V,\gamma}(x)\delta^3(\vec{x} - \vec{y}) \tag{1.12}$$

$$[j_o^{V,\alpha}(x), j_\mu^{A,\beta}(y)]_{x_o=y_o} = [j_o^{A,\alpha}(x), j_\mu^{V,\beta}(y)]_{x_o=y_o} = i\, f_{\alpha\beta\gamma} j_\mu^{A,\gamma}(x)\delta^3(\vec{x} - \vec{y}) \tag{1.13}$$

$$[j_o^{A,\alpha}(x), j_\mu^{A,\beta}(y)]_{x_o=y_o} = i\, f_{\alpha\beta\gamma} j_\mu^{V,\gamma}(x)\delta^3(\vec{x} - \vec{y}) \tag{1.14}$$

Integrating over space variables, we get back the previous commutation relations. Moreover, Gell-Mann postulates the equal time current-current commutators in eqs. (1.12) to (1.14) as fundamental, irrespective of the possible existence of a quark field from which these have been deduced. It should be remarked at this point that these commutators may give rise to additional terms proportional to gradients of the Dirac δ-function in the space variable (called Schwinger terms). We shall say more about this point later on.

Finally, let me summarize briefly the technique used to derive relations between measurable hadronic amplitudes from PCAC, eq. (1.8), and the equal time current-current commutators, eqs. (1.12) to (1.14). The starting point is the Fourier transform of the retarded commutator of two currents taken between hadron states $|a\rangle$ and $|b\rangle$,

$$T^{\alpha\beta}_{\mu\nu} = i \int d^4x \; e^{iqx} \langle b | [j^{\alpha}_{\mu}(x), j^{\beta}_{\nu}(o)] | a \rangle \theta(x_o) \qquad (1.15)$$

where j^{α}_{μ} stands for either vector or axial current. We then evaluate $q^{\mu}T^{\alpha\beta}_{\mu\nu}$ by integration by parts to obtain

$$q^{\mu}T^{\alpha\beta}_{\mu\nu} = - \int d^4x \; e^{iqx} \{ \langle b | [\partial^{\mu} j^{\alpha}_{\mu}(x), j^{\beta}_{\nu}(o)] | a \rangle \theta(x_o) +$$

$$+ \langle b | [j^{\alpha}_{o}(x), j^{\beta}_{\nu}(o)] | a \rangle \delta(x_o) \} , \qquad (1.16)$$

where we dropped the surface term.

The second term on the right hand side of eq. (1.16) is an equal time commutator which we know how to evaluate from eqs. (1.12) to (1.14), while the first term is either zero if j^{α} represents a vector current, or if j^{α} is an axial current, it can be related via PCAC to the production amplitude $\langle b, M^{\alpha} | j^{\beta}_{\nu} | a \rangle$ of a pseudoscalar meson M^{α} (see fig. 1.1).

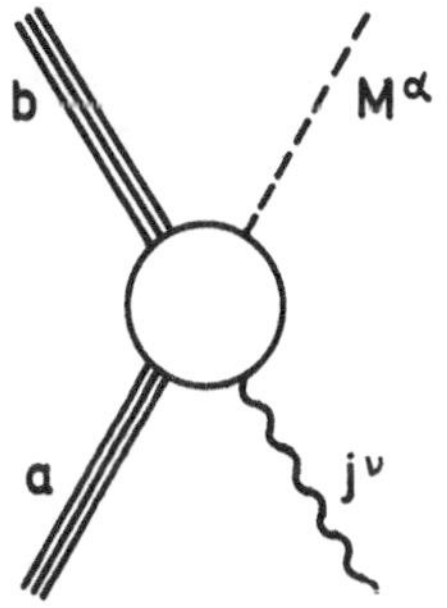

Fig. 1.1

We have

$$\langle b, M^{\alpha} | j^{\beta}_{\nu} | a \rangle = T^{\alpha\beta}_{\nu} =$$

$$= i \int d^4x \; e^{iqx} (\Box + \mu^2) \langle b | [\phi^{\alpha}(x), j^{\beta}_{\nu}(o)] | a \rangle \theta(x_o) \quad (1.17)$$

After substituting the PCAC hypothesis, eq. (1.8), and continuing analytically to $q^2 \neq \mu^2$, we have

$$\tilde{T}^{\alpha\beta}_{\nu} = i \frac{(q^2-\mu^2)}{C} \int d^4x \, e^{iqx} <b|[\partial^{\mu} j^{A,\alpha}_{\mu}(x), j^{\beta}_{\nu}(o)]|a> \theta(x_o) \tag{1.18}$$

which is proportional to the first term in the r.h.s. of eq. (1.16).

There remains the generally untractable term $q^{\mu}T^{\alpha\beta}_{\mu\nu}$ on the left-hand side of equation (1.16). The usual procedure is then to consider the limit $q \to 0$ of equation (1.16), in which case only the poles of $T^{\alpha\beta}_{\mu\nu}$ need to be evaluated, and the residues are generally related to measurable form factors. Combining these results, we obtain the expression

$$\lim_{q \to o} \left(q^{\mu}T^{\alpha\beta}_{\mu\nu} - \frac{iC}{\mu^2} \tilde{T}^{\alpha\beta}_{\nu}\right) = - i \, f_{\alpha\beta\gamma} <b|j^{\gamma}_{\nu}|a> \quad . \tag{1.19}$$

The usefulness of eq. (1.19) depends on whether the off-mass shell amplitude $\tilde{T}^{\alpha\beta}$ at $q^2 = 0$ approximates the corresponding physical amplitude at $q^2 = \mu^2$. The assumption that this is the case is the content of the PCAC hypothesis.

An elementary example which illustrates the application of eq. (1.19) is the relation obtained by Callan and Treiman [3] between the amplitudes for the K-decay processes

$$K^+ \to \pi^o + \ell^+ + \nu \qquad \text{and} \qquad K^+ \to \ell^+ + \nu \quad .$$

The corresponding decay amplitudes are proportional to the current matrix elements

$$<\pi^o|j^{V,-}_{\mu}|K^+> = f_+(k + q)_{\mu} + f_-(k - q)_{u} \tag{1.20}$$

and

$$<0|j_\mu^{A,-}|K^+> = f_K \, k_\mu \qquad (1.21)$$

respectively. In this case we let the hadron states $|a>$ and $|b>$ in eqs. (1.15) to (1.19) be the K^+-meson and the vacuum respectively, and for the currents in the commutator we take the axial current $j_\mu^{A,3}$ and the charged vector current

$$j_\mu^{V,-} = j_\mu^{V,1} - i \, j_\mu^{V,2} \quad .$$

In the limit $q \to 0$ $T_{\mu\nu}^{\alpha\beta}$ has no poles in this case, and we obtain the relation

$$f_+ + f_- = \frac{\mu^2 f_K}{iC} \qquad (1.22)$$

Finally we apply PCAC to the decay amplitude for $\pi^+ \to \ell^+ + \nu$

$$<0|j_\mu^{A,-}|\pi^+> = f_\pi \, q_\mu$$

and obtain

$$C = -i \, f_\pi \mu^2 \qquad (1.23)$$

which combined with eq. (1.22) leads to the result

$$f_+ + f_- = \frac{f_K}{f_\pi} \qquad (1.24)$$

Note that f_+ and f_- are extrapolated here to $q_\mu = 0$. We can compare eq.(1.24) with experiment if we assume it is approximately valid for the physical amplitudes f_+ and f_- at $q^2 = \mu^2$. Unfortunately, at present there are conflicting experimental values of the parameter f_-/f_+ which range somewhere between ± 1.5 ; agreement with eq.(1.24) implies $f_-/f_+ \sim 1/2$.

Of course you know that the first, successful application of current algebra was the famous Adler-Weisberger relation [4] for the renormalization of the axial vector coupling; however we must refer you to the literature for details on this and other interesting applications.

II. Photoproduction of Pseudoscalar Mesons and the Gauge Condition

We consider now the application of current algebra to the photoproduction of pseudoscalar mesons. This example will serve to introduce some of the ideas concerning the electromagnetic and weak interaction contributions to the divergence of hadron currents which will be developed later on.

The amplitude T_μ for the photoproduction of a π^- from a neutron (see Fig. 2.1) takes the form

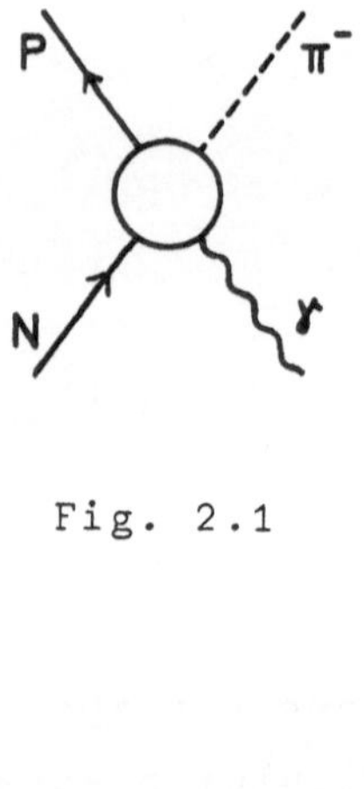

Fig. 2.1

$$T_\mu = \langle \pi^- P | j_\mu | N \rangle =$$

$$= i \int d^4x \, e^{iqx} (\Box + \mu^2) \langle P | [\phi(x), j_\mu(o)] | N \rangle \theta(x_o) \tag{2.1}$$

where

$$j_\mu = j_\mu^{V,3} + 1/\sqrt{3} \;\; j_\mu^{V,8}$$

is the electromagnetic current for the hadrons, and $\phi(x)$ is the charged meson field. The relevant initial and final hadron states in $T^{\alpha\beta}_{\mu\nu}$, eqs. (1.15) to (1.19), are now nucleon states, and the corresponding currents in the commutator are axial and vector currents respectively.

The analytically continued pion photo-production amplitude which enters in eq. (1.19) is then given by

$$\tilde{T}^{\alpha\beta}_{\mu} = i \frac{(q^2-\mu^2)}{C} \int d^4x\, e^{iqx} \times$$

$$\times <P|[\partial^{\nu} j^{A,\alpha}_{\nu}(x), j^{V,\beta}_{\mu}(o)]\theta(x_o)|N> \qquad (2.2)$$

from which we get the photoproduction amplitude, eq. (2.1),

$$T_{\mu} = \lim_{q^2\to\mu^2} \frac{1}{\sqrt{2}} \{\tilde{T}^{1,3}_{\mu} + i\, \tilde{T}^{2,3}_{\mu} + \frac{1}{\sqrt{3}} (\tilde{T}^{1,8}_{\mu} + i\tilde{T}^{2,8}_{\mu})\} . \qquad (2.3)$$

Considering the limit $q \to 0$ of eq. (1.19), we find that $T^{\alpha\beta}_{\mu\nu}$ as well as $\tilde{T}^{\alpha\beta}_{\mu\nu}$ have poles in this case. We will not discuss these pole terms, but refer you to the literature for the relevant details and the relations between electroproduction amplitudes and form factors that are obtained in this limit [5] , [6].

The fundamental point that we want to consider here is the gauge condition on the photoproduction amplitude T_{μ} . From equation (2.1) we have

$$T_{\mu} k^{\mu} = i<\pi^- P|\partial^{\mu} j_{\mu}|N> \qquad (2.4)$$

where $k = p + q - n$ is the photon momentum. Assuming conservation of the electromagnetic current of the hadrons

$$\partial^{\mu} j_{\mu} = 0 \qquad (2.5)$$

we arrive at the gauge condition

$$T_\mu k^\mu = 0 \quad . \tag{2.6}$$

The important thing to notice, however, is that the gauge condition equation (2.6) does not apply to the continued amplitude $\tilde{T}_\mu$ when $q^2 \neq \mu^2$. To see this directly, we turn to equation (1.15) and evaluate $T^{\alpha\beta}_{\mu\nu} k^\nu$. Integrating by parts, we get this time

$$T^{\alpha\beta}_{\mu\nu}\, k^\nu = - \int d^4x\, e^{iqx} \{ <p'|[j^{A,\alpha}_\mu(x), \partial^\nu j^{V,\beta}_\nu(o)]|p> \theta(x_o) + $$
$$+ <p'|[j^{A,\alpha}_\mu(x),\, j^{V,\beta}_o(o)]|p> \delta(x_o) \} \tag{2.7}$$

Substituting for the equal time current-current commutator equation (1.13) and assuming the conservation of the vector current, $\partial^\nu j^{V,\beta}_\nu = 0$ we obtain

$$T^{\alpha\beta}_{\mu\nu}\, k^\nu = - i\, f_{\alpha\beta\gamma} <p'|j^{A,\gamma}_\mu|p> \quad . \tag{2.8}$$

Finally, multiplying equation (1.19) by k^ν, eq. (2.8) by q^μ and taking the difference to eliminate the unknown amplitude $q^\mu T_{\mu\nu} k^\nu$, we obtain [6],[8]

$$\frac{C\, \tilde{T}^{\alpha\beta}_\nu k^\nu}{q^2 - \mu^2} = - i\, f_{\alpha\beta\gamma} <p'|\partial^\nu j^{A,\gamma}_\nu|p> \tag{2.9}$$

This shows that in general $\tilde{T}^{\alpha\beta}_\nu k^\nu = 0$ only if $q^2 = \mu^2$. If we substitute in eq. (2.9) the PCAC hypothesis, eq. (1.8), and the equation of motion for ϕ, $(\Box + \mu^2)\phi^\alpha = j^\alpha$ where j^α (the source function) serves to define the renormalized pion nucleon coupling constant g, we can rewrite equation (2.9) in the form

$$\frac{\tilde{T}^{\alpha\beta}_\nu k^\nu}{q^2 - \mu^2} = i\, f_{\alpha\beta\gamma}\, g K(\Delta^2)\, \frac{\bar{u}(p')\gamma_5 \tau^\gamma u(p)}{\Delta^2 - \mu^2} \tag{2.10}$$

where $\Delta = p' - p$ and $K(\Delta^2)$ is the pion nucleon form factor, normalized to $K(\mu^2) = 1$.

What is the meaning of this result? A simple physical interpretation can be obtained by considering a process involving off-mass shell pions [6]. Since the pion is in fact unstable, let us include a phenomenological interaction for its decay,

$$\mathcal{H}_W = f_\pi \, \phi^- \partial^\mu \, j_\mu^{\ell,+} + \text{h.c.} \tag{2.11}$$

where $j_\mu^{\ell,+}$ is the charged weak lepton current

$$j_\mu^{\ell,+} = \bar{\psi}_\ell \, \gamma_\mu (1 - i\gamma_5) \psi_\nu \quad . \tag{2.12}$$

Due to gauge invariance we must include also an additional electromagnetic coupling

$$\mathcal{H} = i \, e \, f_\pi \phi^- \, j_\mu^{\ell,+} \, A^\mu + \text{h.c.} \quad , \tag{2.13}$$

where A^μ is the photon field.

We now write the amplitude for the photoproduction of a charged lepton-neutrino pair to first order in the weak interaction as a sum of three terms corresponding to the diagrams shown in figs.2.2 and 2.3. The con-

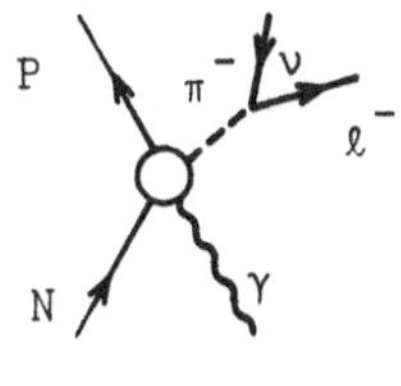

Fig.2.2

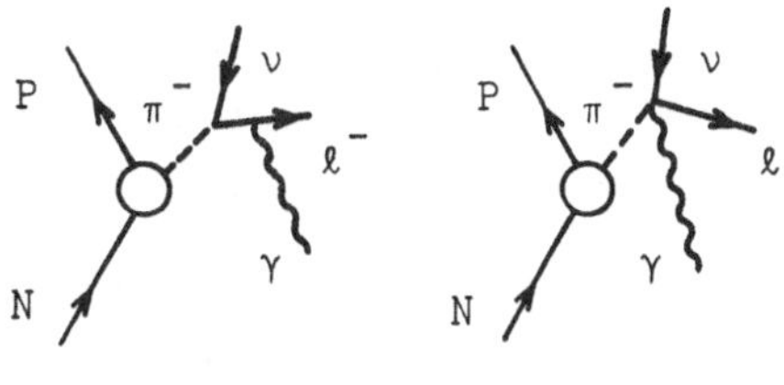

Fig. 2.3

tribution from the electromagnetic interaction of the hadrons is

$$\langle \ell\,\bar{\nu}\,p|j_\mu|n\rangle = f_\pi \bar{u}(q_\ell)\not{q}(1-i\gamma_5)v(q_\nu)\frac{1}{q^2-\mu^2}\tilde{T}_\mu \quad (2.14)$$

where j_μ is the hadron electromagnetic current, while the additional contribution due to the electromagnetic interaction of the charged lepton and the coupling given by equation (2.13) is

$$f_\pi \bar{u}(q_e)\{\gamma_\mu(\frac{1}{\not{q}_\ell-\not{k}-m_\ell})\not{q}(1-i\gamma_5) +$$

$$+ \gamma_\mu(1-i\gamma_5)\}\,v(q_\nu)\,\frac{1}{\Delta^2-\mu^2}\sqrt{2}\,gK(\Delta^2)\bar{u}(p)\gamma_5 u(n) \quad (2.15)$$

Multiplying equations (2.14) and (2.15) by the photon momentum k^μ and setting the sum equal to zero we arrive at the gauge condition

$$\frac{\tilde{T}_\mu k^\mu}{q^2-\mu^2} = \frac{\sqrt{2}\,gK(\Delta^2)}{\Delta^2-\mu^2}\,\bar{u}(p)\,\gamma_5\tau^+\,u(n) \quad . \quad (2.16)$$

This is precisely the result obtained by current algebra, equation (2.10), for the appropriate combination of indices α corresponding to the photoproduction of a π^-. In other words, the equal time commutator between the vector and the axial vector current, eq. (1.13) together with current conservation, leads to the correct gauge condition for the off-mass shell pion photoproduction amplitude. Can we turn the argument around and derive equal time current-current commutators from the gauge condition? We will in fact show that this can be done in the next lecture. To prepare the ground for this demonstration we will derive the above results in yet a third way by considering the divergence of the electromagnetic current j_μ of the hadrons in the presence of the weak interaction, equation (2.11). From the equations of motion for ϕ including the effect of $\mathcal{H}_W$, eq. (2.11), we find

$$\partial^{\mu} j_{\mu} = i\, f_{\pi}\, \phi^{-} \partial^{\mu}\, \bar{\psi}_{\ell} \gamma_{\mu} (1 - i\gamma_{5}) \psi_{\nu} + \text{h.c.} \qquad (2.17)$$

Hence

$$k^{\mu} \langle \ell\nu p | j_{\mu} | n \rangle = - i \langle \ell\nu p | \partial^{\mu} j_{\mu} | n \rangle =$$

$$= f_{\pi} \bar{u}(q_{\ell}) \not{q} (1 - i\gamma_{5}) v(q_{\nu}) \langle p | \phi^{-} | n \rangle \qquad (2.18)$$

and substituting equation (2.14) we get once again eq. (2.16).

A final remark may be appropriate. The so-called gauge condition for off-mass shell amplitudes given in equations (2.9) or (2.16) is a generalized version of the Ward-Takahashi identity [7], which relates a process involving an external photon line to the corresponding process where the photon is absent. You will recall that to derive this identity both the conservation of electromagnetic current $\partial^{\mu} j_{\mu} = 0$ as well as the equal time commutator of the time component of j_{μ} with the charged field $\phi(x)$,

$$[j_{o}(\vec{x},t),\ \phi(\vec{y},t)] = \phi(\vec{x},t)\ \delta^{3}(\vec{x} - \vec{y})$$

are needed. This commutator can be deduced directly from the vector and axial current equal time commutator, eq. (1.13), and the PCAC hypothesis eq. (1.8). The connection between current algebra and generalized Ward-Takahashi identities has been discussed by K. Raman and E. C. G. Sudarshan [8].

III. Divergence Equations

In the previous lecture we discussed the gauge condition or generalized Ward-Takahashi identity for off-mass shell meson photoproduction. This condition can be obtained from either current algebra or by including a weak interaction contribution to the divergence of the electromagnetic current of the hadrons. We will now explore the connection between these two approaches in further detail.

Let us return to photoproduction, but this time consider the electromagnetic contribution to the divergence of the charged axial current, as was done by Adler and Dothan [9]. The principle of minimal electromagnetic interactions leads to the following correction to PCAC:

$$\partial^{\mu} j_{\mu}^{A,+} = C\phi^{-} - i\ e\ j_{\mu}^{A,+}\ A^{\mu} \tag{3.1}$$

Let us introduce the matrix element of $j_{\mu}^{A,+}$ between an arbitrary hadron state containing a single photon $|a\gamma\rangle$ and another hadron state $|b\rangle$,

$$M_{\mu} = \langle b|j_{\mu}^{A,+}|a\gamma\rangle \tag{3.2}$$

Setting the momentum transfer equal to q, we obtain from eq. (3.1) to first order in e,

$$q^{\mu}M_{\mu} = iC\langle b|\phi^{-}|a\gamma\rangle + e\langle b|j_{\mu}^{A,+}|a\rangle \tag{3.3}$$

This expression has essentially the same form as the fundamental equation which we obtained in our discussion of current algebra, eq. (1.16), and it can be used to determine the equal time commutator between the electromagnetic and the axial vector current. To show how this comes about, we reduce the photon in M_{μ}, with momentum k and polarization ε^{ν}.

Then

$$M_\mu = i\int d^4x\ e^{-ikx}\ \Box\langle b|[j_\mu^{A,+}(o),A_\nu(x)]|a\rangle\theta(-x_o)\varepsilon^\nu \tag{3.4}$$

From the equation of motion of the photon field A_ν

$$\Box A_\nu = e\ j_\nu \tag{3.5}$$

and the assumption that $j_\mu^{A,+}$ and A_ν commute at equal time,

$$[j_\mu^{A,+}(x),\ A_\nu(y)\ ]_{x_o=y_o} = 0 \quad , \tag{3.6}$$

we then find

$$M_\mu = \{eT_{\mu\nu}+i\int d^4x\ e^{-ikx}\langle b|[j_\mu^{A,+}(o),\frac{\partial A_\nu(x)}{\partial x_o}]|a\rangle\delta(x_o)\}\varepsilon^\nu \tag{3.7}$$

where $T_{\mu\nu}$ is the appropriate linear combination of amplitudes $T_{\mu\nu}^{\alpha\beta}$ defined in eq. (1.15). Similarly

$$\langle b|\phi^-|a\gamma) =$$

$$= ie\int d^4x\ e^{-ikx}\langle b|[\phi^-(o),j_\nu(x)]|a\rangle\theta(-x_o)\varepsilon^\nu \tag{3.8}$$

where we have set

$$[\phi(o),A_\nu(x)\ ]_{x_o=0} = 0 \tag{3.9}$$

in accordance with eq. (3.6) and PCAC, and assumed for simplicity also

$$[\phi(o),\ \frac{\partial A_\nu(x)}{\partial x_o}] = 0$$

(however this is not essential).

Combining eqs. (3.3), (3.7), (3.8) and (1.16) we arrive at the result

$$\int d^4x\, e^{iqx}[j_o^{A,+}(x), j_\nu(o)]\delta(x_o) =$$

$$= j_\nu^{A,+}(o) + \frac{1}{e}\int d^4x\, e^{iqx}\frac{\partial}{\partial x^\mu}[j_\mu^{A,+}(x), \frac{\partial A_\nu(o)}{\partial x_o}]\delta(x_o) \qquad (3.10)$$

where we have dropped the arbitrary hadron states and the photon polarization vector.
This is precisely the commutation relation between vector and axial vector currents, eq. (1.13), apart from the possible gradient term which may be present if the equal time commutator appearing in eq. (3.7) does not vanish. The important point that we want to emphasize here is that such gradient terms, whether they exist or not, can be ignored when considering relations between physical processes. For example in the photoproduction of a lepton-neutrino pair the physical amplitude is M_μ, eq. (3.2), which is not equal to $eT_{\mu\nu}\varepsilon^\nu$, unless the gradient terms are absent.

Another way to see the connection between the divergence condition, eq. (3.1) and the equal time commutator, eq. (3.10), is to start with the vanishing of the equal time commutator of $j_\mu^{A,+}$ and A_ν, eq. (3.6), and apply to it the D'Alembertian operator. After substituting the divergence condition eq. (3.1), the equation of motion for A_μ, and the canonical commutation relations for $A_\mu(x)$, we get again the result given in eq. (3.10)[10].

An important generalization due to Veltman [11] is to include weak interaction contributions to the divergence of the axial vector currents, eq. (3.1). This can be carried out by introducing charged weak bosons (so far undiscovered experimentally) in analogy with the electromagnetic field, or by introducing a direct weak coupling between the hadron and the lepton currents. It should be evident that this generalization gives us also the results of equal time commutation relations between axial currents, eq. (1.14); for example, the Adler Weisberger relation.

To complete this discussion, let me show you how one obtains the electromagnetic and weak interaction contributions to the divergence of the currents from an interaction Lagrangian and the equal time current-current commutators. To be specific let me consider the contribution of the electromagnetic interaction to the divergence of the charged strangeness conserving vector current $j_\mu^{V,+}$ in theories without gradient couplings,

$$\mathcal{L}_{em} = e\, j_\mu A^\mu \tag{3.11}$$

From Lorentz invariance we have

$$\partial^\mu j_\mu^{V,+}(x) = i\,[P^\mu, j_\mu^{V,+}(x)] \tag{3.12}$$

where P^μ is the four momentum. Since isospin is a good symmetry apart from the electromagnetic interaction, the only nonvanishing part of the commutator, eq. (3.12), comes from the electromagnetic interaction contribution, eq. (3.11), to the energy operator P^0 [12]. Hence we can write eq. (3.12) in the form

$$\partial^\mu j_\mu^{V,+}(\vec{x},t) = -ie\int d^3y\,[j_\mu(\vec{y},t)A^\mu(\vec{y},t), j_0^{V,+}(\vec{x},t)] \tag{3.13}$$

Substituting the equal time commutator between vector currents eq. (1.13) we obtain

$$\partial^\mu j_\mu^{V,+}(x) = -\,ie\; j_\mu^{V,+}(x)A^\mu(x) \tag{3.14}$$

An alternative approach is to consider the usual definition of the currents from the variation of the total Lagrangian due to physe transformations of the fields produced by the unitary operators of the symmetry group. In this case the divergence of these currents is proportional to the variation of the non-symmetric part of the Lagrangian [13]; in our example this is the electromagnetic interaction, eq. (3.11). One readily finds

$$\partial^{\mu} j^{\alpha}_{\mu}(x) = - i \left[\mathcal{L}_I(x), J^{\alpha}(t) \right]_{x_o = t} \tag{3.15}$$

provided the interaction does not contain gradients, and substituting eq. (3.11), we get again eq. (3.14).

Finally there is of course always the pedestrian approach of assuming a field theory model, quarks for example, and computing directly the divergence of the currents from the field equations.

We have now come around a full circle and shown how to obtain the contribution of electromagnetic and weak interactions to the divergence of the currents by using the equal time current-current commutators. This completes the proof of equivalence between the divergence and the current algebra approach to lowest order in the electromagnetic and weak interaction couplings. An interesting question which remains to be investigated is the possibility that in processes involving second order electromagnetic interactions, for example η decay, these two methods may lead to quite different results.

References

1. M. Gell-Mann and M. Lévy, Nuovo Cim. 16, 105 (1960).
2. M. Gell-Mann, Phys. Rev. 125, 1067 (1962); Physics 1, 63 (1964).
3. C. Callan and S. B. Treiman, Phys. Rev. Lett. 16, 153 (1966).

4. S. L. Adler, Phys. Rev. Lett. 14, 1051 (1965); Phys. Rev. 143, B736 (1965); W. Weisberger, Phys. Rev. Lett. 14, 1407 (1965); Phys. Rev. 143, 1302 (1966).
5. S. L. Adler, Argonne Proc.Inter. Conf. on Weak Interactions, Oct. 1965.
6. M. Nauenberg, Phys. Lett. 22, 201 (1966).
7. Y. Takahashi, Nuovo Cim. 6, 37 (1957).
8. K. Raman and E. C. G. Sudarshan, Phys. Lett. 21, 450 (1966).
9. S. L. Adler and Y. Dothan, Phys. Rev. 151, 1267 (1966).
10. M. Nauenberg, Phys. Rev. 154, 1455 (1967).
11. M. Veltman, Phys. Rev. Lett. 17, 553 (1966).
12. J. S. Bell, CERN TH 725, Nov. 1966.
13. I am indebted to Professor S. Berman for many useful discussions on this point.

ALGEBRAIC FORMULATION OF DYNAMICAL PROBLEMS†

By

P. BUDINI

International Centre for Theoretical Physics
Trieste, Italy

Introduction

The extension of Wigner supermultiplet theory of isospin to unitary symmetry has led to the consideration of SU(6) as the fundamental symmetry group for hadron physics [1]. By this extension, such a phenomenon as the nuclear anomalous magnetic moment found simple and elegant explanation, despite the expectation that these phenomena could only be understood in the frame of orthodox dynamical theory. This success encouraged the thought that a further extension of the symmetry to include the Poincaré group would have given the basis for a fundamental dynamical theory of strong interactions [2].

It is known that the numerous attempts at relativistic extension of SU(6) have been only partially successful, and certainly one of the main difficulties encountered in such attempts is connected with the fact that any theory with a symmetry containing the Poincaré

---

† Lecture given at the VI. Internationalen Universitätswochen f. Kernphysik, Schladming, 26 February-11 March 1967.

group admits automatically complete mass degeneracy [3]; the mass splitting of the physical particles is obtained only after breaking the symmetry. An unsatisfactory procedure indeed, to get what should be one of the main dynamical results of the theory from a breaking of the theory itself.

In order to avoid this basic difficulty it was suggested that groups be introduced which from the beginning allowed mass splitting. These groups were not symmetry groups for the system but, while containing these as subgroups, were capable of generating the right energy eigenstates of the physical system. Such groups were subsequently called dynamical groups [4], non-invariance groups or spectrum-generating groups [5], and should have been suitable to formulate the fundamental dynamical properties of the system in algebraic form. We will call them briefly spectral groups (or algebras).

Obviously, since this was an attempt at a new methodology, which had to show its power on systems with no classical analogue, it had first to be tested on exactly soluble quantum-mechanical systems. Of these only the non-relativistic systems possess self-consistent satisfactory solutions which can be reached without introducing the complications of quantum field theory. We will show first how the method works for these systems. We will then describe how old and recent attempts at relativistic theories can be formulated in the frame of this methodology.

The Method

The whole field is still in the development stage so it is certainly not possible to talk of a theory or of an established method. One can only put down what could be a tentative sketch of a method.

a) Let S be a symmetry algebra of a system defined by the property to be the largest algebra whose generators commute with the mass operator M of the system. Then in the space where the generators of S build up a complete set, M is a function of the Casimir operators of S and the eigenstates belonging to the different eigenvalues of M form the basis for irreducible representations of S.

b) We will suppose then M (or a function of it) to be a member of a larger algebra SG (compact or not) containing S as a maximum compact subalgebra, and to be able to find a representation of SG which is the direct sum of the irreducible representations of S belonging to the eigenvalues of M, each representation of S being contained in that of SG at most once.

c) Then the Casimir operator of SG which will contain both the mass operator M (or a function of it) and the Casimir operator of S, may furnish the functional dependence of the former on the latter and constitute the basis for the dynamical solution of the problem. The Casimir operator of S can further be expressed in the form of a differential operator on an appropriate symmetric space and will give the dynamical wave equation of the systems in this space.

In the proposed method two interconnected ambiguities may arise. First, the definition of M. It is clear that the definition of M will depend on the system described and will in general be unambiguous only for systems with classical analogues. For relativistic systems M must be equal to P_μ^2 where P_μ are the generators of the translation of the centre of mass. So this ambiguity should arise only for non-relativistic systems with no classical analogue.

Second, the identification of spectral algebra SG. It will be seen in the examples given that more than one algebra SG may be chosen. But provided they satisfy the criteria given in b) they allow the determination of the dependence of M from the Casimir operator of SG. The essential difference between them will be that, in general, they generate different parts of the spectrum of the system.

The Harmonic Oscillator

One of the best examples of application of the proposed method is the isotropic oscillator. It is known that the n-dimensional isotropic harmonic oscillator possesses a symmetry represented by the real unitary group SU(n) [6]. In fact its Hamiltonian is (we put $m = 1$, $\omega = 1$, $h = 1$) :

$$H = \sum_{1}^{n} (q_j^2 + p_j^2) = \frac{1}{2} \sum_{1}^{n} \{a_j^+, a_j\} \tag{1}$$

where the operators a_j, a_j^+ are defined by

$$a_j = \frac{1}{2}(p_j + iq_j) \quad ,$$

$$a_j^+ = \frac{1}{2}(p_j - iq_j)$$

and obey the commutation relations

$$[a_i, a_j^+] = \delta_{ij}$$

the other commutators vanishing. (We will always indicate by [] the commutator and by {} the anticommutator). The operators

$$E_i^j = \frac{1}{2} \{a_i, a_j^+\} \tag{2}$$

are hermitian and obey the commutation relations

$$[E_j^i, E_\ell^k] = \delta_{jk} E_\ell^i - \delta_{i\ell} E_j^k \tag{3}$$

and build up the algebra U(n). This algebra can be decomposed in SU(n) × U(1) where U(1) consists of the generator

$$H = \sum_{1}^{n} E_i^i$$

Besides, H commutes with E_j^i; therefore SU(n) is the symmetry algebra of the n-dimensional harmonic oscillator.

The eigenvectors belonging to the eigenvalues of H will then build up the basis for irreducible representation of SU(n). The general eigenvector is

$$\frac{1}{\sqrt{\alpha_1!\ldots\alpha_n!}} \; a_1^{+\alpha_1}\ldots a_n^{+\alpha_n} \, |0> \qquad (4)$$

where the state $|0>$ is defined by

$$a_i |0> = 0$$

and the irreducible representations are the completely symmetric representations of SU(n) corresponding to one row Young diagram or the generalized triangular weight diagram in n-1 dimensions. We have now to find an algebra containing SU(n) as a subalgebra and having a representation as direct sum of symmetric representations of SU(n) based on the space of all vectors (4).

At least two such algebras exist [7]. One is the non-compact algebra SP(n,R) defined by the generators (2) and

$$E_o^{ij} = \frac{1}{2} \{a_i^+, a_j^+\}$$

$$E_o^o = H$$

$$E_{ij}^o = -\frac{1}{2} \{a_i , a_j\} \qquad (5)$$

with commutators

$$[E^i_j , E^{\ell k}_o] = \delta_{jk} E^{i\ell}_o + \delta_{j\ell} E^{ik}_o$$

$$[E^i_j , E^o_{k\ell}] = -\delta_{ik} E^o_{j\ell} - \delta_{i\ell} E^o_{jk}$$

$$[E^{k\ell}_o , E^o_{mn}] = \delta_{nk} E^\ell_m + \delta_{n\ell} E^k_m + \delta_{m\ell} E^k_n + \delta_{mk} E^\ell_n$$

$$[E^o_o , E^{k\ell}_o] = 2 E^{k\ell}_o$$

$$[E^o_o , E^o_{k\ell}] = - 2 E^o_{k\ell}$$

$$[E^{k\ell}_o , E^{mn}_o] = [E^o_{k\ell} , E^o_{mn}] = 0 \qquad (6)$$

It is easy to show that the bilinear Casimir operator of Sp(n,R) can be expressed in the following form [8]:

$$Q_2 = B_2 + \frac{H^2}{2} + \frac{1}{2} \{E^o_{ij} , E^{ij}_o\} \qquad (7)$$

The value of Q_2 is fixed once we fix the representation of Sp(n,R) as direct sum of the symmetric representations of SU(n):

$$Q_2 = - \frac{n}{2}(n + \frac{1}{2}) \qquad (8)$$

B_2 is the Casimir operator of SU(n) which for the symmetric representations is

$$B_2 = \frac{n-1}{n} s(s+n) \qquad (9)$$

where all representations belonging to the H-eigenstates are obtained for s = 1, 2, ... ($s = \alpha_1 + \ldots + \alpha_n$).

The anticommutators in (7) can be calculated explicitly since the representation is fixed and we have

$$\frac{1}{2} \{E^o_{ij} , E^{ij}_o\} = - H^2 - \frac{n}{2} (\frac{n}{2} + 1) \qquad (10)$$

Substituting (10) in (7) we obtain the required dependence of H on the Casimir operator B_2 of SU(n):

$$H^2 = \frac{n}{n-1}\left[B_2 + \frac{n}{2}\left(\frac{n}{2} - \frac{1}{2}\right)\right] \tag{11}$$

and substituting (9)

$$H^2 = \left(s + \frac{n}{2}\right)^2, \quad s = 0, 1, 2, \ldots \tag{12}$$

It has to be pointed out that to exhaust all representations of the oscillator one has to take two infinite irreducible representations of Sp(n,R) (or one of its universal covering group). One is the sum of all symmetric representations of SU(n) with s = 0, 2, 4,... and the other with s = 1, 3, 5,... They can be distinguished by the eigenvalue of the parity operator P defined by

$$P\, a_i = -\, a_i$$

$$P\, a_i^+ = -\, a_i^+ \tag{13}$$

which are +1 and -1 for the first and second respectively. (P commutes both with H and Sp(n,R)). They are both defined by the same eigenvalue of Q_2 . It can be shown that the relations derived and in particular (12) can be deduced from the commutation relations (3) and (6) independently from the particular realization of the algebras once the representation space is fixed.

From (11) a wave equation for the stationary states can be obtained by expressing B_2 as a differential operator (Laplace-Beltrami) in the symmetric space of rank 1,

$$\frac{Sp(n,R)}{SU(n) \times U(1)}$$

The interesting feature of such an equation is that the

potential is substituted by the symmetry property of the space. For example in the case of n = 2 one obtains from (11):

$$H^2\psi = (4\Delta_\Omega + 1)\psi$$

where

$$\Delta_\Omega = -\frac{\partial^2}{\partial\beta^2} - \cot\beta\frac{\partial}{\partial\beta} - \frac{1}{\sin^2\beta}\left(\frac{\partial^2}{\partial\alpha^2} + \frac{\partial^2}{\partial\gamma^2}\right) + \frac{2\cos\beta}{\sin^2\beta}\frac{\partial^2}{\partial\gamma^2}$$

and α, β, γ are the Euler angles which parametrize the SU(2)-symmetric space. Eigenfunctions of this equation are the matrix elements $D^{(j)}_{mm'}(\alpha\beta\gamma)$ of the (2j+1)-dimensional irreducible representation of SU(2) corresponding to the eigenvalues (12) with s = 2j.

The second algebra we want to discuss briefly is the non-compact SU(n,1).

In this case we take the generators (2) and Ref. [7]

$$E^i_o = g(H)\; a^+_i$$

$$E^o_i = f(H)\; a_i$$

$$E^o_o = -H + c \tag{14}$$

which close the algebra SU(n,1) provided*

$$H + c = \eta f(H) f^+(H) + \tfrac{1}{2}$$

$$H + c = \eta g(H) g^+(H) - \tfrac{1}{2}$$

$$(E^o_i)^+ = \eta E^i_o \tag{15}$$

for η = -1 (or to the algebra SU(n + 1) for η = 1)**.

* The constant c can be set to zero taking traceless generators for E^i_j.

** In this case representations of SU(n+1) exist which contain all the symmetric representations of SU(n) up to a maximum value of n [7].

The Casimir operator Q_2 of the larger group can be expressed again as a function of H and the Casimir operator B_2 of the symmetry group; independently of the functions f and g because of (15) [9]:

$$Q_2 = B_2 - \frac{n-1}{n} H^2 + \frac{n(2n-1)}{4(n+1)} \tag{16}$$

For

$$Q_2 = - \frac{n^2(n+2)}{4\,(n+1)} \tag{17}$$

one obtains a representation which is the direct sum of all symmetric representations of SU(n). Substituting (17) and (9) in (16) one obtains (12). It is to be noted that in this case one single representation of SU(n,1) contains all the symmetric representations of SU(n) (the essential generators of SU(n,1) do not commute with the parity operator).

In conclusion it is seen by this example that different spectrum generating algebras can be obtained, which generate different parts of the physical spectrum of the system. They have in common the property that their Casimir operator defines the dependence of the Hamiltonian on the Casimir operator of the symmetry group.

The Hydrogen Atom

It has long been known [10] that the Schrödinger theory of the hydrogen atom possesses a symmetry represented by the group O(4). The components of angular momentum M_i and Runge-Lenz vector A_i

$$\vec{A} = \frac{1}{2\mu}(\vec{p} \times \vec{M} - \vec{M} \times \vec{p}) - e^2 \frac{\vec{r}}{r}$$

commute in fact with

$$P_o = \frac{p^2}{2\mu} + \frac{e^i}{r}$$

and obey the commutation relations

$$[\vec{M} , \vec{M}] = i \vec{M}$$

$$[\vec{A} , \vec{A}] = - 2i \frac{P_o}{e^4\mu} \vec{M}$$

$$[\vec{M} , \vec{A}] = i \vec{A} \qquad (18)$$

but do not build up a closed algebra which can be obtained only after multiplying A_i by $(-e^4\mu/2P_o)^{1/2}$:

$$\vec{N} = (\frac{-e^4\mu}{2P_o})^{1/2} \vec{A}$$

$\vec{M}$ and $\vec{N}$ do in fact build up the algebra of O(4) for P_o negative c-number (for P_o positive the algebra of the Lorentz group). The symmetry then is sui generis: valid only in every multiplet subspace, and the generators depend, in every multiplet subspace, on P_o.

The irreducible representations of O(4) are characterized by the two invariants

$$F = \frac{1}{2}(M^2 + N^2)$$

$$G = \vec{M} \cdot \vec{N} \qquad (19)$$

The hydrogen atom corresponds to the representation D(j,j), G = 0, F = 4j(j+1) and each P_o level is $(2j+1)^2$ times degenerate. The fact that this is the representation realized is due to the "classical" fact that the Runge-Lenz vector $\vec{N}$ lies in the plane of the orbit and is perpendicular to the angular momentum vector $\vec{M}$, thus $G = \vec{M} \cdot \vec{N} = 0$.

In order to make evident the SO(4) symmetry of the hydrogen atom-Schrödinger equation, Fock [10] has projected the p space on a four dimensional unit sphere of unit radius introducing the variables on the sphere

$$\xi_i = \frac{2p_o p_i}{p_o^2 + p^2}$$

$$\xi_4 = \frac{p_o^2 - p^2}{p_o^2 + p^2} \qquad (20)$$

where $p_o = \sqrt{-2\mu P_o}$. One has $\xi^2 = 1$.

The Schrödinger equation in integral form becomes

$$\Phi(\xi) = \frac{\mu e^4}{p_o \pi^2} \int \frac{\Phi(\xi')}{(\xi-\xi')^2} \delta(\xi'^2-1)\, d^4\xi' \qquad (21)$$

with

$$\Phi(\xi) = \frac{1}{4} p_o^{-5/2} (p_o^2 + p^2)^2 \phi(p) \qquad (22)$$

and the SO(4) symmetry is evident.

It was pointed out by Fock that the eigensolutions of (21) are the four-dimensional spherical harmonics, and the eigenvalues are those of Balmer.

It was shown [11] that the spectral algebra of the non-relativistic hydrogen atom is SO(4,1). That is, an irreducible representation of this algebra exists which is the direct sum of the SO(4) representations belonging to the hydrogen atom energy levels.

In order to obtain the generators of the SO(4,1)algebra one can project the four-dimensional sphere $\xi^2=1$ on a five-dimensional hyperboloid $\pi^2 - \pi_5^2 = 1$:

$$\pi_\mu = \frac{2c\, \xi_\mu}{c^2 - \xi^2} \quad ; \quad \pi_5 = \frac{c^2 + \xi^2}{c^2 - \xi^2}$$

the generators of the SO(4,1) algebra

$$U_{\alpha\beta} = - i \left(\pi_\alpha \frac{\partial}{\partial \pi_\beta} - \pi_\beta \frac{\partial}{\partial \pi_\alpha}\right) \qquad \alpha,\beta = 1,2,\ldots 5$$

written in terms of ξ variables, become (for $\xi^2 = c^2 = 1$) :

$$V_{\mu\nu} = - i \left(\xi_\mu \frac{\partial}{\partial \xi_\nu} - \xi_\nu \frac{\partial}{\partial \xi_\mu}\right) \tag{23}$$

$$V_{5\mu} = - i \left[\frac{\partial}{\partial \xi_\mu} - \xi_\mu \left(\xi_\nu \frac{\partial}{\partial \xi_\nu} + 2\right)\right] \tag{24}$$

The representations of SO(4,1) are characterized by the invariants

$$q = \frac{1}{2} V_{\mu\nu} V_{\mu\nu} - V_{5\mu} V_{5\mu} \tag{25}$$

and

$$W = - W_\alpha W^\alpha \tag{26}$$

with

$$W^\alpha = \frac{1}{8} \varepsilon^{\alpha\beta\gamma\delta\lambda} V_{\beta\gamma} V_{\delta\lambda}$$

W can be expressed in the form

$$W = G . R$$

where G is the biquadratic invariant of the O(4) subgroup and F depends on $V_{5\mu}$. Since G is zero in the multiplet subspaces

$$W = 0 \tag{26'}$$

Correspondingly the energy depends on only one quantum number, as it should. There are two classes of infinite unitary representations for W = 0 : those with $q > 0$

and those with $q = -(n+1)(n+2) \leq 0$ (n integer) and they can all be expressed in terms of the irreducible representations of O(4) spanned by the spherical harmonics on the $\xi^2 = 1$ sphere. On this space q can be easily calculated:

$$q = \frac{1}{2}(\xi_\mu \frac{\partial}{\partial\xi_\nu} - \xi_\nu \frac{\partial}{\partial\xi_\mu})^2 + \frac{\partial^2}{\partial\xi_\mu{}^2} - (\xi_\mu \frac{\partial}{\partial\xi_\mu})^2 - $$
$$- 2(\xi_\mu \frac{\partial}{\partial\xi_\mu}) - 4 = -4 \qquad (27)$$

and the irreducible representation is completely determined. The space of the representation is the direct sum of O(4) representation subspaces,each subspace being contained at most once. For the invariants given by (26') and (27) the space is characterized by:

$$\psi = \sum_{j=o}^{\infty} \oplus \psi_{j+1}(\xi) \qquad (28)$$

Matrix elements of the generators (23) and (24) can be easily obtained by the usual procedure.

From the Casimir operator q one can obtain in this case also both the energy spectrum and the wave equation on the symmetric space of the system. We have in fact:

$$- \frac{1}{2} V_{\mu\nu} V_{\mu\nu} \equiv 1 + \frac{e^4 \mu}{2P_o} \qquad (29)$$

and it is easy to check that the non-compact part of the invariant q is also diagonal in every multiplet subspace as anticipated:

$$V_{5\mu} V_{5\mu} = (j+1)^2 + 3 \qquad (30)$$

Substituting (30) and (29) in (25) one obtains the Balmer formula

$$P_o = - \frac{e^4\mu}{2(j+1)^2} \qquad (31)$$

Further, from (26) one obtains the equation

$$(\Delta_\Omega + 1)\ \psi(\xi) = -\frac{e^4\mu}{2P_o}\ \psi(\xi) \tag{32}$$

where Δ_Ω is the angular part of the four-dimensional Laplacian. Eq. (32) is equivalent to the ordinary Schrödinger equation for the hydrogen atom. The new feature is that having written the equation on the four-dimensional sphere $\xi^2 = 1$ space of symmetry for SO(4), the potential term does not appear.

One can now try to write the Schrödinger equation directly in π space, taking into account that the π now act as operators in the space of the SO(4,1) representation. We will not give the details of the calculation, which are identical with those of Fronsdal [13]. (The transformation taken by Fronsdal is $\pi_\mu = \xi_\mu \pi_5$ which brings the $\xi^2 = 1$ sphere on the hypercone $\pi^2 - \pi_5^2 = 0$ and is equivalent with ours in the limit $c^2 = \xi^2 = 1$.) The result is

$$\left[\pi_o\left(E + \frac{p_o^2}{2\mu}\right) + \pi_4\left(E - \frac{p_o^2}{2\mu}\right) + e^2 p_o\right]\ \psi(\xi) = 0$$

where π_o and π_4 are now operators (in the present case infinite matrices). It is to be pointed out that ψ is now a one-column infinite matrix containing all eigenstates of the system; and the eigensolution of this equation no longer depends on p_o, which is now just a parameter. This form of the hydrogen atom-Schrödinger equation brings us near to the relativistic equations.

Lorentz-Invariant Equations

Let us now consider relativistic systems. We assume that the behaviour of any system, composite or not, or more precisely of its centre of mass, be described, when the system is isolated, by an equation of the type

$$(i\gamma^{\mu} \frac{\partial}{\partial x_{\mu}} + \kappa)\ \psi(x) = 0 \qquad (33)$$

$$\mu = 0, 1, 2, 3 \quad \text{metric } g_{oo} = -g_{11} = -g_{22} = -g_{33} = 1$$

where κ is a Lorentz scalar.

If $L \equiv |\alpha_{\mu}^{\nu}|$ is a (proper) Lorentz transformation, the corresponding linear operator T_L to act on the function $\psi(x)$ is defined by

$$\psi'(x') = T_L\ \psi(x)$$

where $x' = Lx$. It is easily shown then that the necessary and sufficient condition for eq. (33) to be Lorentz covariant (for $\kappa \neq 0$) is

$$T_L\ \gamma^{\mu}\ T_L^{-1} \alpha_{\mu}^{\nu} = \gamma^{\nu} . \qquad (34)$$

If we now go to the Lorentz algebra with generators $M_{\nu\mu}$:

$$[M_{\mu\nu}, M_{\rho\sigma}] = -i(g_{\mu\rho}M_{\nu\sigma} + g_{\nu\sigma}M_{\mu\rho} - g_{\mu\sigma}M_{\nu\rho} - g_{\nu\rho}M_{\mu\sigma}) \qquad (35)$$

and invariants

$$2\ F = \frac{1}{4}\ \{M_{\mu\nu}M^{\mu\nu}\} = \frac{1}{2}(j_o^2 + \lambda^2 - 1)$$

$$G = \varepsilon^{\mu\nu\rho\sigma}\ M_{\mu\nu}\ M^{\rho\sigma} = i\ j_o\lambda \qquad (35')$$

condition (34) becomes

$$[M_{\mu\nu}, \gamma_{\rho}] = i(g_{\nu\rho}\gamma_{\mu} - g_{\mu\rho}\gamma_{\nu}) \quad . \qquad (34')$$

It is known that equation (33) can be derived by an invariant Lagrangian density provided an invariant bilinear product $(\psi_1 \psi_2)$ can be defined* .

We will preferably use eq. (33) written in momentum space (plane wave solutions) :

$$(p^\mu \gamma_\mu - \kappa)\, \psi(p) = 0 \qquad (33')$$

In this case $p^2 = p_o^2 - \vec{p}^2$ is by definition the square of the system total energy and the ambiguity of the Hamiltonian definition of the non-relativistic case disappears.

It can be shown (J. M. Gel'fand) [14] that from the commutation relations (34') some but not all features of the matrices γ_μ can be determined. Precisely because the operator p_μ behaves as a Lorentz vector, it is obvious that given one irreducible representation of the Lorentz group $D(j_o\lambda)$ spanned by ψ , $p_\mu\psi$ will in general decompose into four interlocking representations, and the commutation relations (34') determine the matrix elements of γ_μ as Clebsch-Gordan coefficients which couple the interlocking representations; but the choice of the representations which are interlocking is left free. And this choice determines the various kinds of eq. (33). It is found [14] that if $\tau \equiv$ $\equiv (j_o, \lambda)$ and $\tau' \equiv (j_o', \lambda')$ are the interlocking representations, the matrix elements of γ_o are $\neq 0$ only if $(j_o', \lambda') = (j_o \pm 1, \lambda)$ or $(j_o', \lambda') = (j_o, \lambda \pm 1)$ and in these cases

$$\gamma_{oj}^{(\tau,\tau')} = c^{(\tau,\tau')} \sqrt{(j+j_o+1)(j-j_o)} \qquad (36)$$

---

* If one requires further invariance against the complete Lorentz-group = proper + space reflection, the condition $[S, \gamma_o] = 0$ must be added where S is the generator of space reflections.

or

$$\gamma_{oj}^{(\tau,\tau')} = c^{(\tau,\tau')} \sqrt{(j + \lambda + 1)(j - \lambda)} \quad \text{respectively}$$

with $c^{(\tau,\tau')}$ arbitrary, and the other matrices γ_μ can be determined from (34').

The fact that the space spanned by the components of the ψ decomposes into a direct sum of spaces where irreducible representations of the Lorentz group act explains why eq. (33) is suitable to represent multimass systems. In fact the space spanned by ψ can then also be decomposed into subspaces belonging to several irreducible representations of the Poincaré group and O'Raifeartaigh's theorem [3] does not apply.

For this reason eq. (33) is the ideal equation to start with in order to fulfil our programme.

We will now show how the classification of the various possible eqs. (33) and their solution can be performed with the use of algebraic methods. Let us suppose that ψ has a finite number of components (as in the equations of Dirac, Pauli-Fierz, Kemmer etc.), then the successive application on them of the operators will bring from one to the other of the irreducible Lorentz subspaces but the process will have an end or be brought to the starting point. That is, the space spanned by ψ will in general be a basis for a bigger algebra of which the γ_μ are elements.

In fact for all equations considered we will define

$$[\gamma_\mu, \gamma_\nu] = - S_{\mu\nu} \tag{37}$$

It is easy to check that, because of (36), we have:

$$[M_{\mu\nu}, S_{\sigma\rho}] = i(g_{\mu\rho} S_{\nu\sigma} + g_{\nu\sigma} S_{\mu\rho} - g_{\mu\sigma} S_{\nu\rho} - g_{\nu\rho} S_{\mu\sigma}) \tag{38}$$

In case $S_{\mu\nu}$ build up a representation of the Lorentz algebra, i.e., they obey commutation relations (35), the $S_{\mu\nu}$ and γ_μ build up the algebra SO(3,2) [15] (it can be shown that this case corresponds to the separability of orbital and spin terms in $M_{\mu\nu}$). This will happen in the case of the equations considered in the following and the equations will in fact be classified according to the different representations of this algebra which are spanned by the wave function ψ and we will show precisely how to classify them through the Casimir operators of SO(3,2).

We will consider two classes:

A) κ is a number

In order to compare the present case with the nonrelativistic one, let us write eq. (33') in the form

$$p^o \gamma_o \psi = (\bar{p} \cdot \bar{\gamma} + \kappa) \qquad (33'')$$

We see that the energy operator in the space spanned by ψ has the symmetry of the operator $\bar{p} \cdot \bar{\gamma} + \kappa$, which is that of the maximal compact subgroup SO(3) of SO(3,1). So it is expected that the energy eigenvalues will depend on the quantum number j which characterizes the SO(3) Casimir operator. But eq. (33") is SO(3,1)-invariant so that a single column eigenfunction will generally contain all energy eigenstates of the system. Moreover, since we have that the equation has to represent a relativistic multimass system, the space spanned by ψ will be the basis for a reducible SO(3,1) representation. In fact, in the cases considered it will span an irreducible SO(3,2) representation and we will classify the equations according to their SO(3,2) content.

Summarizing, we have that the sequence SO(3) - SO(3,1) - SO(3,2) represents the symmetry algebra and the spectral algebras respectively, and the equation has the invariance of the spectral algebras.

The Casimir operators of SO(3,2) are

$$2\,Q = \frac{1}{4}\{S^{\mu\nu}S_{\mu\nu}\} + \frac{1}{2}\{\gamma_\mu,\gamma^\mu\}$$

$$= \frac{1}{4}\{S^{ab}S_{ab}\}$$

$$R = w^a w_a \qquad (39)$$

where $w^a = \varepsilon^{abcde}\beta_{bc}\beta_{de}$ $a,b,\ldots = 0,1,\ldots 4$, where, taking into account that the γ_μ with the $S_{\mu\nu}$ build up the algebra SO(3,2), we have put

$$\gamma_\mu = S_{4\mu}$$

Let us now examine in particular the two extreme cases of the Dirac and Majorana equations:

1) Dirac equations

In this case the space spanned by the wave function is minimum. The interlocking representations of SO(3,1) are the conjugate ones (1/2, 3/2) and (1/2,-3/2) with Casimir operators

$$F_L = 3/4,\ G = \pm\, 3/4$$

The Casimir operators of the unique representation of SO(3,2) are Q = 5/4, R = 9/8. The system has only two energy states. Castell [15] has given the weight diagram of the representation appearing in the equations of Dirac, Kemmer and Bhabha. We will go directly to the

2) Majorana equation

This is the opposite case in which the space spanned by ψ is infinite. Majorana [16] took it as a basis for a unitary infinite irreducible representation of the Lorentz group represented by (0,1/2) or (1/2,0).

Equivalently, one could consider it spanned by interlocking representations: but the choice of interlocking representations is not uniquely determined in this case, and we will see again that this indeterminacy does not alter the dynamical content of the equation; the fact is that they all belong to the same representation of SO(3,2).

The matrix element of γ_o take the form

$$\gamma_o^{(\tau,\tau')} = c(j + \tfrac{1}{2})\ \delta_{jj'}\ \delta_{mm'} \tag{40}$$

We have further that the γ_μ and $S_{\mu\nu}$ obey the anti-commutation relation

$$\{S^\mu_\nu\ ,\ S^\rho_\mu\} + \{\gamma_\nu,\gamma^\rho\} = -\delta^\rho_\nu \quad . \tag{41}$$

From it, putting $\nu = \rho$, summing and taking (39) into account we have

$$Q = \tfrac{1}{8}\{S^{ab}S_{ab}\} = -\ \tfrac{1}{2} + \tfrac{1}{2}\{\gamma_\nu,\gamma^\nu\} = -\ (F_L + 1) \tag{42}$$

which establishes an important relation between the Casimir operators of the SO(3,2) and SO(3,1) representations realized in the Majorana equation. We obtain easily

$$F_L = -\ \tfrac{3}{8}\ ,\quad G = 0\ ;\quad Q = -\ \tfrac{5}{8}\ ,\ R = 0 \tag{43}$$

We will now use eq. (41) to determine the energy spectrum of the system described by the Majorana equation.

Let us multiply (33') by the operator $(p_\mu\gamma^\mu+\kappa)$ and, taking into account (41), we have

$$(\gamma^\mu p_\mu + \kappa)(\gamma_\mu p^\mu - \kappa)\psi = (-p^2-\kappa^2-\ \tfrac{1}{2}p_\nu p^\rho\{S^\nu_\mu,S^\mu_\rho\})\psi = 0\ .$$

Let us consider the operator

$$W = \frac{1}{4} p_\mu p^\mu \{S_{\mu\nu}, S^{\mu\nu}\} - \frac{1}{2} \{S_\mu^\nu, S_\rho^\mu\} p_\nu p^\rho \quad . \qquad (44)$$

It is easy to check that it is an invariant of the Poincaré group; in fact it commutes both with $M_{\mu\nu}$ and with p_μ (this is due simply to the invariance of eq. (33) which determines (34) and (38)). The equation becomes then:

$$[-\frac{1}{2} p^2 - \kappa^2 - \frac{1}{4} p^2 \{S_{\mu\nu}, S^{\mu\nu}\} + W] \psi =$$

$$= (\frac{1}{4} p^2 + W - \kappa^2) \psi = 0 \qquad (45)$$

where we have taken into account that

$$\frac{1}{4} \{S_{\mu\nu}, S^{\mu\nu}\} = 2 F_L = - 2Q - 2 = - \frac{3}{4} \quad .$$

We know that the space spanned by ψ can be broken up into subspaces of definite j with 2j+1 components (cf. eq. (40)). In these spaces W is diagonal and has the value $p^2\bar{M}^2 = p^2 j(j+1)$. (45) becomes then

$$[\frac{1}{4} p^2 + p^2 j(j+1) - \kappa^2] \psi_{j,m} = 0 \quad , \qquad (46)$$

which gives the mass spectrum

$$p^2 = \frac{\kappa^2}{(j + \frac{1}{2})^2} \qquad (47)$$

(It must be noted that the negative energy solution which has been added in the multiplication by $(\gamma^\mu p_\mu + \kappa)$ has to be discarded because γ_o has only positive eigenvalues).

Similarly one can classify and solve intermediate cases as those of Kemmer, Bhaba, etc.

B) κ is a Lorentz-invariant operator

Let us take

$$\kappa = - p_4 \kappa_4 - \kappa_o \tag{48}$$

With p_4 a number and γ_4 operator, (33') becomes

$$(p^a \gamma_a - \kappa_o)\psi(p) = 0, \qquad a = 0,1,2,3,4 \tag{49}$$

with κ_o constant. An equation of this form has been proposed by Nambu [17] and discussed by Fronsdal [13]. p_4 is considered to build up together with p_μ a vector in a five-dimensional space with metric

$$g_{00} = - g_{11} = - g_{22} = -g_{33} = - g_{44} = 1 \ .$$

Let β_{ab} be the generators of a SO(4,1) group of transformations in this space. Then eq. (49) is invariant for these transformations provided the γ satisfy the corresponding five-dimensional relation of eq. (34'). Eq. (49) can be put in the form (33") where on the right-hand side we have $\bar{p} \cdot \bar{\gamma}$ substituted by a four-dimensional scalar product. We have then again that the symmetry algebra will be SO(4) and the energy eigenvalues will depend on the Casimir operators of this group. And again SO(4,1), SO(4,2) will be the sequence of the spectral algebras. The equation is SO(4,1)-invariant and so a single ψ will contain all the energy eigenstates of the system.

We can then repeat, extended to five dimensions, the arguments and methods used in case A.

We will again consider the case when the $\beta_{ab} = [\gamma_a, \gamma_b]$ and γ_a close up the algebra of SO(4,2).

Of the various possibilities inside this frame let us consider the correspondence of the Majorana one. $\psi(p)$ will span the space of a unitary irreducible representation of SO(4,1). We have then that the β and γ obey the anticommutation relation

$$\{\beta^a_b , \beta^c_a\} + \{\gamma_b , \gamma^c\} = -2\delta^c_b \tag{50}$$

Summing over c = b we have again for the first Casimir operator Q of SO(4,2):

$$Q = \frac{1}{10}\{\beta^{AB} , \beta_{AB}\} = \frac{1}{10}[\{\beta^{ab} , \beta_{ab}\} + 2\{\gamma^c , \gamma_c\}] =$$

$$= \frac{1}{10}[2q + 2\{\gamma^c , \gamma_c\}] = -2 - \frac{2}{10}q \tag{51}$$

where q is the Casimir operator of SO(4,1) and the indices A,B run from 0 to 5 .

In order to find the mass spectrum of eq. (49) let us apply to it the operator $(p_a \gamma^a + \kappa_o)$; we obtain

$$[-(p^2 - p_4^2) - \frac{1}{2} p^a p_b \{\beta^c_a , \beta^b_c\} - \kappa_o^2]\psi = 0 . \tag{52}$$

We have that the invariants of the non-homogeneous SO(4,1) group are

$$P = p^2 - p_4^2$$

$$W = \frac{1}{2} p_a p^a \beta^{cd}\beta_{cd} - \frac{1}{2}\{\beta^d_a , \beta^b_d\} p^a p_b . \tag{53}$$

It can be easily checked (e.g., going to the frame $(p_o, 0000)$) that the value of the invariant W is

$$W = (p^2 - p_4^2)\Omega$$

where Ω is the quadratic invariant of the SO(4) subalgegra. If we consider that representation of SO(4,1) which is the direct sum of SO(4) most degenerate representations, we have

$$\Omega = n^2 - 1$$

Substituting the anticommutator in (52) from (53) we obtain for every n^2 multiplet subspace

$$\{(p^2 - p_4^2)(n^2 - 2) - \frac{q}{2}(p^2 - p_4^2) - \kappa_o^2\}\psi_{n\ell} = 0 \quad (52')$$

which gives the spectrum

$$p^2 = p_4^2 + \frac{\kappa_o^2}{n^2-2-\frac{q}{2}}$$

It is worth noting that in order to get the spectrum to depend on only one quantum number, q must have a unique value. This is in fact the case; for the Majorana type spectrum, which is the same as that of the hydrogen atom (cf. eq. (27)), we have q = -4 and

$$p^2 = p_4^2 + \frac{\kappa_o^2}{n^2} \quad ,$$

as found by Nambu. For non-unitary representations one obtains by the same method

$$p^2 = p_4^2 - \frac{\kappa_o^2}{n^2}$$

For further discussions about the properties of this and similar types of equations we refer to the papers of Nambu and Fronsdal [13], [17].

Another possible generalization of eq. (33') with κ-invariant operator is

$$[p^\mu\gamma_\mu - \kappa_o\,(\frac{W}{p^2} + \frac{1}{4})] = 0 \quad (54)$$

where W is the biquadratic Poincaré invariant operator (44). We can take ψ to span the unitary irreducible representation of the Lorentz group identical with the

Majorana ones: $(0,\frac{1}{2})$ or $(\frac{1}{2},0)$.

Multiplying this equation by the operator

$$[p_\nu\gamma^\nu + \kappa_o(\frac{W}{p^2} + \frac{1}{4})]$$

and with the same procedure used in the Majorana case we obtain the mass spectrum:

$$p^2 = \kappa_o^2 \, (j + \frac{1}{2})^2$$

for both representations $(0,\frac{1}{2})$ and $(\frac{1}{2},0)$, this means for both integer and half integer j. This spectrum has some interesting similarity with the elementary particle mass spectrum where baryon and meson masses seem to have the same spin dependence despite the fact that they obey different statistics.

In eq. (54), integer and half integer spin solution can be handled separately (for example by imposing baryon conservation) since they belong to different Lorentz representations, and in the quantized theory the spin and statistics difficulty should not arise.

Conclusion

We have shown how the algebraic method can be employed to solve some dynamical problems and the important role that is played, in this frame, by the Casimir operators. In particular, whilst in the non-relativistic case the Casimir operator of the spectral algebra gives the connection between the mass operator and the Casimir operator of the symmetry algebra, in the relativistic cases considered one has still a one-to-one dependence of the Casimir operator of the spectral algebra on that of the symmetry algebra, but this dependence is no longer sufficient to give the mass spectrum which is obtained by introducing the Casimir operator of the inhomogeneous spectral algebra which leads to an equation of the Klein-Gordon type, giving the mass spectrum implied by the equation. In this way the whole procedure is kept at the algebraic level.

One point which in our opinion deserves further attention is the connection between the properties of the higher algebras and the corresponding symmetric spaces. It would be interesting to see if one can always represent the whole dynamical behaviour of a system (the whole spectrum) in the frame of a single equation going to a higher dimensional space which bears the symmetry of the higher algebra embracing that of symmetry, and if the internal structure of relativistic systems, composite or not, can always be represented by inserting into equations of the type (33) appropriate operators for the γ and κ.

References

1. F. Gürsey and L. Radicati, Phys. Rev. Lett. 13, 299 (1964).
2. R. Delbourgo, A. Salam and J. Strathdee, Proc. Roy. Soc. A 284, 146 (1965); M. A. Bég and A. Pais, Phys. Rev. Lett. 14, 404 (1965); L. Michel and B. Sakita, Ann. Inst. Henri Poincaré 2, 167 (1965);P. Budini and C. Fronsdal, Phys. Rev. Lett. 14, 968 (1965).
3. L. O'Raifeartaigh, Phys. Rev. Lett. 14, 332 and 575 (1965); Phys. Rev. 139, 1052 (1965).
4. A. O. Barut, Phys. Rev. 135, B 839 (1964); A. O. Barut and A. Böhm, Phys. Rev. 139, 1107 (1965).
5. Y. Dothan, M. Gell-Mann and Y. Ne'eman, Phys. Lett. 17, 148 (1965).
6. J. M. Jauch and E. L. Hill, Phys. Rev. 57, 641(1940).

7. R. C. Hwa and J. Nuyts, Phys. Rev. 145, 1188(1966).
8. G. Bisiacchi and P. Budini, Nuovo Cim. 44, 418 (1966).
9. G. Bisiacchi and T. Chersi, Thesis,to be published in Nuovo Cim.
10. W. Pauli, Z. Phys. 36, 336 (1926); V. Fock, Z. Phys. 98, 145 (1935); V. Bargmann, Z. Phys. 99, 576 (1936).
11. A. O. Barut, P. Budini and C. Fronsdal, Proc. Roy. Soc. A 291, 106 (1966).
12. P. Budini, Nuovo Cim. 44, 363 (1965).
13. C. Fronsdal, Preprint UCLA, THR 18 (1966).
14. A. M. Yaglom, Zh. Eksperim. i Teor. Fiz. 18, 1105 (1948).
15. J. K. Lubanski, Physica 9, 310 (1942); L. Castell, Z. Naturf. 20a, 737 (1965).
16. E. Majorana, Nuovo Cim. 9, 335 (1932).
17. Y. Nambu, Preprint, EFINS 66-65.

HIGH ENERGY SCATTERING OF HADRONS†

By

J. J. J. KOKKEDEE
CERN - Geneva

In these lectures we shall discuss some recent developments in the theory of elastic scattering of hadrons (strongly interacting particles) at high energy and small momentum transfers. By "high energy" we mean here the region $p_L \gtrsim 4$ GeV/c, where p_L is the laboratory momentum of the incident particle; the highest value of p_L available with present accelerators is about 30 Gev/c for protons. This region is characterized by the fact that cross sections are found to be smooth, slowly varying, in certain cases even nearly constant, functions of p_L without any structure. By "small momentum transfers" we mean the region of the forward diffraction peak. At present, experimental high energy data are available for PP, PN, $\bar{P}P$, $\bar{P}N$, $\pi^{\pm}P$, $K^{\pm}P$ and $K^{\pm}N$ reactions, where P = proton, N = neutron. We shall deal with two recently proposed models which have been rather successful in predicting relations between the various meson-baryon and baryon-baryon cross sections at high energies, namely the quark model [1] - [8] (Ch. II), and the combination of Regge pole model and current algebra [9], [10] (Ch. III). In Ch. I, we briefly mention a few general

† Lecture given at the VI. Internationalen Universitätswochen f. Kernphysik, Schladming, 26 February-11 March 1967.

asymptotic properties of elastic scattering amplitudes, which are useful for the discussion in the last two Chapters.

I. Generalities

1. Asymptotic expansion of the amplitude

Consider the elastic scattering process (A and B are hadrons*)

$$A + B \to A + B \quad , \tag{1}$$

together with the crossed reaction

$$\bar{A} + B \to \bar{A} + B \quad , \tag{2}$$

in the region of small momentum transfers ($0 \lesssim |t| \lesssim 1$ $(\mathrm{GeV}/c)^2$) and high energies ($s \gtrsim 8\ \mathrm{GeV}^2$) where $t = -(\text{c.m. momentum transfer})^2$ and $s = -(\text{c.m. energy})^2$. In this region we write down the following asymptotic series expansions for the scattering amplitudes $T(s,t)$ and $\bar{T}(s,t)$ for reactions (1) and (2) respectively

$$T(s,t) \underset{s\to\infty}{\simeq} \sum_{j=o,1,\dots} c_j(t)\, s_t^{\alpha_j(t)} \quad , \tag{3}$$

$$\bar{T}(s,t) \underset{s\to\infty}{\simeq} \sum_{j=o,1,\dots} \bar{c}_j(t)\, s_t^{\alpha_j(t)} \quad , \tag{4}$$

in which $s_t = s + \frac{t}{2} - m_A^2 - m_B^2$. The coefficients c_j and $\bar{c}_j$ can, in general, be complex. The exponents α_j are taken real with the condition $\alpha_o > \alpha_1 > \dots$, as suggested by the fact that no oscillations have been found experimentally in the variation of total cross sections at high energy.

---

* We always neglect spins.

The existing high-energy data can be described in the form (3) and (4) with a small number of terms (two to four). Note that (3) and (4) correspond to dominance by poles in the angular momentum plane; if $\alpha(t)$ varies with t these poles are Regge poles, otherwise they are fixed poles. Logarithmic terms of the form $(\ln s_t)^{\beta_j(t)}$ in (3) and (4), which we have left out because present experiments are not accurate enough to decide on their presence or absence, would reflect the presence of fixed or moving cuts in the l plane.

Using analyticity of T in s, crossing symmetry, and the assumption of slower than exponential growth of T in the upper-half s plane, a strong relation between the two expansions (3) and (4) can be derived [11] - [14], namely

$$\bar{c}_j(t) = c_j^*(t)\, e^{-i\pi\alpha_j(t)} \quad . \tag{5}$$

This important property holds only for the expansions (3) and (4) in which s_t, instead of s, is used as expansion parameter. An immediate consequence of (5) is that these expansions contain actually the same powers of s_t.

Consider the forward direction (t = 0). If the total cross sections for scattering of hadrons tend to a finite non-zero limit at very high energies, i.e., using the optical theorem

$$\sigma_T = \frac{8\pi^2}{k\sqrt{s}} \operatorname{Im} T(s,o) \quad , \tag{7}$$

(k = c.m. momentum, σ_T = total cross section), if $\operatorname{Im} T(s,o) \underset{s\to\infty}{\sim}$ const. s, we have $\alpha_o(o) = 1$ and hence from (5)

$$\operatorname{Im} c_o(o) = \operatorname{Im} \bar{c}_o(o) \quad , \tag{8}$$

$$\operatorname{Re} c_o(o) = - \operatorname{Re} \bar{c}_o(o) \quad . \tag{9}$$

Relation (8) is the Pomeranchuk theorem $\sigma_T(\infty) = \bar{\sigma}_T(\infty)$ for total cross sections [15]. The present data [16], [17] extending up to $p_L \sim 25$ GeV/c are compatible with it although even at 25 GeV/c there are still considerable differences $\bar{\sigma}_T - \sigma_T$. These have to be described by the second and higher terms in the expansions (3) and (4). Relation (9) gives $X(s,0) = -\bar{X}(s,0)$ for $s \to \infty$, where

$$X(s,t) = \frac{\text{Re } T(s,t)}{\text{Im } T(s,t)} \quad .$$

Since for $p_L \gtrsim 5$ GeV/c both $X(s,0)$ and $\bar{X}(s,0)$ are found to be negative (and of order -0.3 for PP and NP, -0.2 for $\pi^{\pm}P$) where measured [18], (9) suggests that $X(\infty,0) = \bar{X}(\infty,0) = 0$, which would mean that the measured values of X and $\bar{X}$ are due to the higher terms in (3) and (4). Note that for an amplitude which has positive exchange signature [13], i.e., which satisfies

$$\bar{T}(s,t) = \xi T(s,t) \quad , \tag{10}$$

with $\xi = +1$, the assumption $\alpha_o(o) = 1$ implies the asymptotic amplitude to be purely imaginary. This applies to the Regge pole model if the Pomeranchuk trajectory passes through one (Ch. III).

Suppose that in addition to the asymptotic term (the one with $j = 0$) only one term ($j = 1$) is needed to describe the data above ~ 5 GeV/c; we then have from (3), (4), (5) and (7) for an imaginary asymptotic amplitude, putting $\Delta = [\sigma_T(s) - \sigma_T(\infty)]k\sqrt{s}/8\pi^2$ and $\alpha_1(0) = \alpha_1$,

$$\Delta = -\cos(\alpha_1\pi)\, \bar{\Delta} - \sin(\alpha_1\pi)\text{Re}\bar{T} \quad , \tag{11a}$$

$$\text{Re}T = \cos(\alpha_1\pi)\text{Re}\bar{T} - \sin(\alpha_1\pi)\, \bar{\Delta} \quad . \tag{11b}$$

The total cross section curves suggest in the high-energy region $\alpha_1 \simeq 1/2$. Since Δ and $\bar{\Delta}$ are found to be positive above 5 GeV/c for all hadrons, (11) predicts for this value of α_1 the real parts to be negative, in accordance with all existing measurements [18]. Moreover, $\Delta < \bar{\Delta}$ implies $|\mathrm{Re}\,\bar{T}|<|\mathrm{Re}\,T|$ for all pairs of reactions (1) and (2). For instance, for PP and $\bar{P}P$ scattering (P = proton), we expect around 10 GeV/c and t = 0, $|\mathrm{Re}\,\bar{T}|<<|\mathrm{Re}\,T|$, because here $\bar{\Delta} >> \Delta$ (the bar refers to $\bar{P}P$) due to strong annihilation contributions in $\bar{P}P$ scattering [14]. This agrees with recent experimental results [19].

We remark that the assumption $\alpha_o(o) = 1$ is not absolutely required, neither by theory nor by experiment. In fact, good fits to the existing cross section data exist with $\alpha_o(o)$ slightly less than one (see Ch. III), implying vanishing asymptotic total cross sections [10].

At non-forward directions ($t \neq 0$, but small) we have at once from (5) in the limit of infinite energies for each pair of processes (1) and (2)

$$|T| = |\bar{T}| \,, \qquad \text{or} \qquad \frac{d\sigma_{el}}{dt} = \frac{d\bar{\sigma}_{el}}{dt} \,. \tag{13}$$

This is the generalized Pomeranchuk theorem [20]. Since all measured differential cross sections for $0 \leq |t| \lesssim 1 (\mathrm{GeV}/c)^2$ can be parametrized in the form

$$\frac{d\sigma_{el}}{dt} = \left(\frac{d\sigma_{el}}{dt}\right)_{t=o} e^{at+bt^2} \,, \tag{14}$$

with the slope a always in the neighbourhood of $10 (\mathrm{GeV}/c)^{-2}$ and $b^2/a<<1$, (13) implies

$$a = \bar{a} \quad , \quad b = \bar{b} \quad , \quad \text{for } s \to \infty \,. \tag{15}$$

The existing data above $\sim$ 5 GeV/c are again compatible with this result [16]. For instance, for $\pi^{\pm}P$ scattering, (15) is well satisfied with constant a and b. The slopes

$\bar{a}$ and a for $\bar{P}P$ and PP show a variation with energy, more pronounced for $\bar{a}$ than for a. The latter seems to increase with increasing energy (shrinking of the diffraction peak), the former decreases (expansion of the diffraction peak), both tending to a common limit of about $10(\mathrm{GeV}/c)^{-2}$, which is about the same value as obtained from $\pi^{\pm}P$ scattering [21]. Around 10 GeV/c, however, a and $\bar{a}$ are still very different [$a \simeq 9(\mathrm{GeV}/c)^{-2}$, $\bar{a} \simeq 12.5(\mathrm{GeV}/c)^{-2}$]. This difference and the expansion of the $\bar{P}P$ diffraction peak are connected with the presence and variation with energy of $\bar{P}P$ annihilation contributions [22]. Similar results are known for K-nucleon scattering, where again the slopes a and $\bar{a}$ tend to common limits. From this we may conclude that the leading term in the expansions (3) and (4) has exponent $\alpha_o(t) = \alpha_o(o)$ for $|t| \lesssim 1(\mathrm{GeV}/c)^2$, with $\alpha_o(o) = 1$, or very close to one.

2. Unitarity condition

We write the unitarity condition in the form

$$\frac{16\pi^2}{k\sqrt{s}} \operatorname{Im} T(s,t) = \sum_n \langle AB|\tau^+|n\rangle\langle n|\tau|AB\rangle \quad , \qquad (16)$$

where the sum extends over all elastic and inelastic final states and where τ is the properly normalized scattering operator. Conservation of energy and momentum is assumed to be taken into account. For t = 0, (16) becomes

$$2\sigma_T(AB) = \sum_n \langle AB|\tau^+|n\rangle\langle n|\tau|AB\rangle \quad . \qquad (17)$$

Consider the reaction

$$A + B \rightarrow \text{inelastic final state } n,$$

in the c.m. system. The jet structure of high energy inelastic collisions (small transverse momenta in c.m.) suggests that the particles in the final state may be roughly divided into two groups A' and B', one containing the particles which go forward in the c.m. system and one containing the particles going backward. In general A' and B' are many-particle systems. In analogy with two-body reactions, we can now, according to the possible exchanges in the t channel in reactions of the type A + B → A' + B', roughly subdivide the contributions to the sum in (17) (the same, of course, holds for (16)), i.e., to $\sigma_T(AB)$, into three classes:

a) annihilation or baryon exchange contributions $\sigma_A(AB)$;
b) charge, isospin and strangeness exchange contributions $\sigma_E(AB)$;
c) Pomeranchuk contributions $\sigma_P(AB)$.

Class a) contributes only when A and B have opposite baryon number. Class c) refers to those contributions for which vanishing internal quantum numbers are exchanged in the t channel of A + B → A' + B'. One expects that at very high energies only the latter contributions survive, i.e.,

$$\frac{\sigma_A}{\sigma_P} \to 0 \quad , \quad \frac{\sigma_E}{\sigma_P} \to 0 \quad , \quad \text{for } s \to \infty \quad , \tag{18}$$

and that at finite energies the measured differences $\sigma_T(\bar{A}B) - \sigma_T(AB)$ are mainly determined by contributions of classes a) and b).

As a consequence of the above subdivision we may write approximately*

$$\sigma_T(AB) = \sigma_P(AB) + \sigma_A(AB) + \sigma_E(AB) \quad . \tag{19}$$

---

* We neglect the fact that in general the three types of contributions affect each other due to absorption corrections. For instance, the presence of σ_A affects σ_{el} and hence σ_P .

The contribution from elastic final states in (17), as being mainly diffractive, is supposed to be included in σ_P. From (19) simple inequalities between total cross-sections follow [5]. For instance,

$$\sigma_T(PN) > \sigma_T(PP) \quad ,$$

which one obtains by noting that $\sigma_P(PP) \simeq \sigma_P(PN), \sigma_A(PP) = \sigma_A(PN) = 0$ and $\sigma_E(PN) > \sigma_E(PP)$, the last inequality arising from the fact that charge can be exchanged in more ways between N and P than between P and P. Similarly, because of strong annihilation contributions in $\bar{P}N$, one has

$$\sigma_T(\bar{P}N) > \sigma_T(PP) \quad .$$

Similar inequalities can be obtained for meson-baryon scattering. Although these considerations are very crude, all inequalities between measured high-energy cross sections derived on the basis of (19) turn out to agree with the facts [5].

II. Quark Model for High-Energy Scattering

1. The additivity assumption

A very successful approach to high-energy scattering of hadrons has recently [1] ÷ [8] been developed on the basis of the quark model [23]. In the latter the hadrons are assumed to be composed of SU(3) triplets, called quarks, which have fractional charge and baryon number. We shall denote the triplet of quarks by p, n, λ, which have the following quantum numbers

$$p : \quad B = \frac{1}{3} , \; Q = \frac{2}{3} e , \; S = 0, \quad I = \frac{1}{2}$$

$$n : \quad B = \frac{1}{3} \, , \; Q = -\frac{1}{3} e \, , \; S = 0 \, , \; I = \frac{1}{2}$$

$$\lambda : \quad B = \frac{1}{3} \, , \; Q = -\frac{1}{3} e \, , \; S = -1 , \; I = 0 \quad . \qquad (20)$$

The corresponding triplet of antiquarks is denoted by $\bar{p}$, $\bar{n}$, $\bar{\lambda}$. In the quark model the baryons and decuplet members have quark composition (q q' q"), the pseudoscalar and vector mesons ($q \, \bar{q}'$). For instance, P = (ppn), $\pi^+ = (p\bar{n})$, $K^+ = (p \, \bar{\lambda})$, etc. In the following we shall always forget about the spin variables; more precisely, we shall consider only spin averaged cross sections [24].

The basic assumption one makes in the application of the quark model to high-energy scattering is the so-called additivity assumption, in which the amplitude for a two-body reaction $A + B \rightarrow A' + B'$ is written as a sum of corresponding two-body quark scattering amplitudes, the quarks inside hadrons being considered as quasi-free particles*. Restricting ourselves in these lectures to elastic scattering amplitudes ($A' = A$, $B' = B$) we can put this assumption into mathematical form in the following way [3], [5]

$$T'_{AB}(s,t) = \sum_{ij} f_i^A(t) \; f_j^B(t) \; T'_{ij}(s,t) \qquad , \qquad (21)$$

where i and j label the quarks and antiquarks contained in A and B respectively. The function f_i^A is the form factor for the i-th quark (or antiquark) in A; with a similar definition for f_j^B. The primes in (21) indicate that the amplitudes are non-covariant amplitudes, i.e.,

* As to the nature of the quarks we are completely ignorand; it may be that quarks exist as truly free particles although they have never been detected as yet; it may also be that they can only exist as a kind of quasi-particles which cannot be separated from the hadrons, like phonons in a crystal; it may also be ...

matrix elements of the S matrix. This means that, in contrast to the amplitudes T and $\bar{T}$ defined in Section I. 1., their normalization is such that at high energy s independence of the amplitudes implies s independence of cross sections. It is expected that (21) is applicable only at high energy ($s \gtrsim 10$ GeV^2) and small momentum transfers [$|t| \lesssim 1(\text{GeV}/c)^2$]. Its main applications up to now have been in the forward direction where the imaginary part of the left-hand side becomes proportional to the total cross section $\sigma_T(AB)$. It should be mentioned that SU(3) need not be an exact symmetry of the S matrix; the only point at which it intervenes in the quark model is in the construction of the bound state wave functions.

The energy dependence of T'_{ij} in (21) deserves a comment. The value of $T'_{ij}(s,t)$ for given s and t and fixed quark labels i and j will, in general, not be the same for a nucleon-nucleon and a meson-nucleon collision, the reason being that the c.m. energy of the $q_i + q_j$ collision will be different for the two cases and unequal to the energy of the A + B collision. If the meson and baryon have the same laboratory momentum p_L, the effective momenta of the constituting quarks should be on the average $\frac{1}{2}p_L$ and $\frac{1}{3}p_L$ respectively*. This implies that in checking relations, obtained from (21) by eliminating quark amplitudes, which involve both meson-baryon and baryon-baryon cross sections, these cross sections should be taken at laboratory momenta in the ratio 2 to 3 respectively [3], [5].

Before applying the additivity assumption to specific processes we have to make an important restriction concerning its applicability to antibaryon-baryon and baryon-baryon reactions [8]. Consider the imaginary

---

* This makes sense only if the effective mass m_q of bound quarks is of order $m_q \approx \frac{1}{3} m_B \approx \frac{1}{2} m_M$, where m_B and m_M are average values of the baryon and meson masses respectively.

parts of the elastic amplitudes together with the unitarity condition (16). If in (16) A = $\bar{B}'$ = antibaryon and B = baryon, at finite energies an appreciable contribution - called class a) in Ch. I - to the sum over n comes from annihilation channels $\bar{B}' + B \rightarrow$ mesons (Fig. 1). Since such a process involves baryon, or

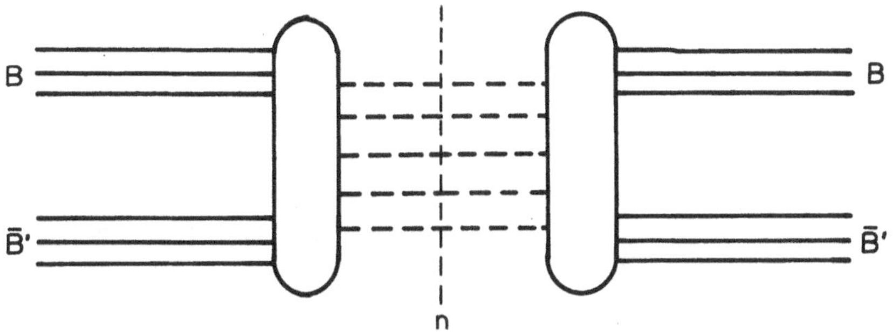

quarks
antiquarks

Fig. 1

triple quark, exchange, it can clearly not be described by the additivity formula. In other words, Fig. 2, which describes (16) under the additivity assumption cannot be a good approximation to the annihilation process of Fig. 1. So $\bar{B}'B$ annihilation contributions have to be excluded from the additivity assumption for the imaginary part of the $\bar{B}'B$ amplitude*; for these contributions an assumption of multiplicative, rather than additive, type to express them in terms of quarks seems more natural (see the end of this Chapter). The exclusion of annihilation processes from the additivity assumption for Im $T_{\bar{B}'B}$ may in general affect also Im $T_{B'B}$, as well as Re $T_{\bar{B}'B}$ and Re $T_{B'B}$, because these amplitudes are rela -

* A similar restriction applies to meson - baryon backward scattering. Its contribution to the sum in (16) for small angle meson-baryon scattering is, however, very small.

ted by the crossing relations (5). Take for example the

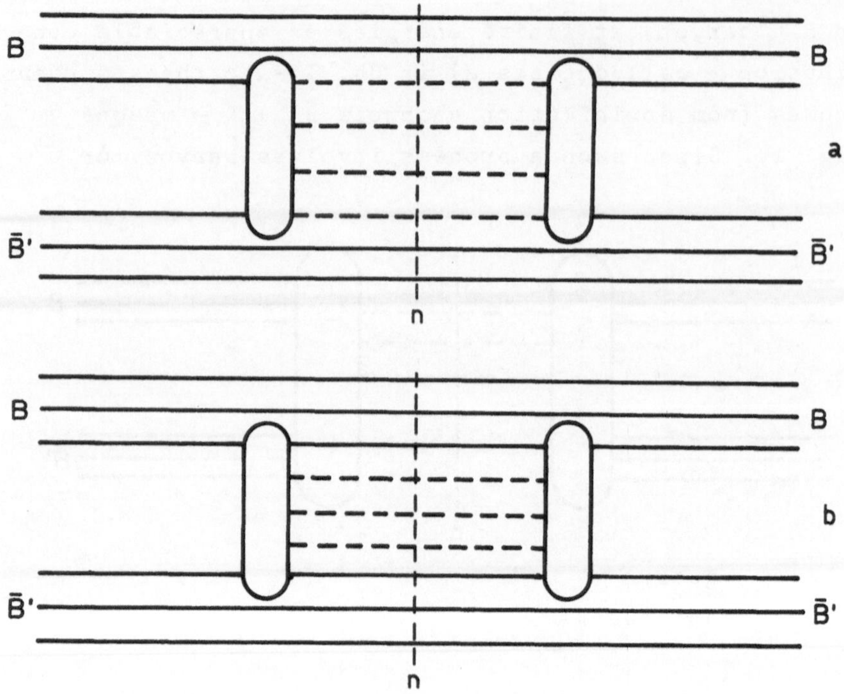

Fig. 2

case of PP and $\bar{P}P$ scattering in the forward direction. If only two terms are necessary to describe the high-energy data we can apply equations (11) to this case (the bar refers to $\bar{P}P$). Present high energy data suggest the value $\alpha_1 \approx \frac{1}{2}$, so that (11) becomes

$$\mathrm{Re}\, T = -\bar{\Delta} \quad , \quad \mathrm{Re}\, \bar{T} = -\Delta \quad .$$

Since $\bar{\Delta} \approx \sigma_A$, where σ_A is the cross section for annihilation (see below), the exclusion of annihilation from the additivity assumption implies in this case the same for the main part of Re T. However, Δ, and therefore Im T, and Re $\bar{T}$ are not affected. This is fortunate since we wish to define additivity for the complete non-annihilation part of the sum in (16), i.e., in the case of PP scattering, for the imaginary part of the complete amplitude. Note that for $\alpha_1 \simeq 0$ we would have an inconsistency. In that case (11) becomes

$$\mathrm{Re}\, T = \mathrm{Re}\, \bar{T} \quad , \quad \Delta = - \bar{\Delta} \ ,$$

so that, assuming additivity to hold for the complete imaginary amplitude Im T, the same would be true for Im$\bar{T}$, including annihilation.

The part to be excluded in the additivity scheme from the sum in (17) for $\sigma_T(\bar{B}'B)$ is just the cross section for annihilation $\sigma_A(\bar{B}'B)$,

$$2\sigma_A(\bar{B}'B) = \sum_{n'} \langle \bar{B}'B|\tau^+|n'\rangle\langle n'|\tau|\bar{B}'B\rangle \quad ,$$

which is an experimental quantity ($|n'\rangle$ contains only mesons). However, only one high-energy annihilation cross section has been measured so far, namely $\sigma_A(\bar{P}P) = 22.5 \pm 2$mb at $p_L = 5.7$ GeV/c [25]. At this momentum one has $\sigma_T(\bar{P}P) = 62.5 \pm 1$ mb and $\sigma_T(PP) = 41.0 \pm 1$ mb so that

$$\sigma_A(\bar{P}P) = [\sigma_T(\bar{P}P) - \sigma_T(PP)] \pm (1 \div 2)\ \mathrm{mb} \quad . \qquad (22)$$

In order to be able to compare our theoretical results with experiment we shall assume that (22) continues to hold for $p_L \gtrsim 6$ GeV/c and that the part of the total cross section $\sigma_T(\bar{B}'B)$ which is additive in the quark amplitudes is $\sigma_T(\bar{B}'B) - \sigma_A(\bar{B}'B)$. As long as accurate data on annihilation cross sections are lacking, a more rigorous procedure which takes care of small corrections due to the fact that annihilation and non-annihilation contributions are not completely independent [22] does not make much sense.

The diagrams of Fig. 2 describe for $B\bar{B}'$ reactions

the two types of contributions that approximate in the additivity scheme the non-annihilation part of (16). Similar diagrams can be drawn for baryon-baryon and meson-baryon scattering. The right-hand half of Fig. 2a contains antiquark-quark annihilation, i.e., single quark exchange between hadrons A and B in $\langle AB|T|n\rangle$. Such inelastic reactions, however, have never been found to occur experimentally. We,therefore, expect that only the diagrams of the type depicted in Fig. 2b and not those of Fig. 2a can provide a good approximation to the non-annihilation part of the sum in (16). Hence, we also exclude antiquark-quark annihilation contributions of the type given by Fig. 2a from the additivity scheme (see also remarks following (28)). Note that if quarks do not exist as separate particles but only in bound combinations [26], it would be a priori clear that a process as shown in Fig. 2a cannot exist.

We now use the additivity assumption, expressed by (21), limited to non-annihilation contributions as discussed above, to derive certain relations between hadron cross sections. We first consider the <u>forward direction</u> ($t = 0$), for which $f_i^A = f_j^B = 1$. We obtain from (21) the following relations

$$\tilde{S}^{\pm}(PP) = 5\,\tilde{S}^{\pm}(pp) + 4\,\tilde{S}^{\pm}(np) \quad , \tag{23a}$$

$$\tilde{S}^{\pm}(PN) = 4\,\tilde{S}^{\pm}(pp) + 5\,\tilde{S}^{\pm}(np) \quad , \tag{23b}$$

$$S^{+}(\pi^{+}P) = 3\,\tilde{S}^{+}(pp) + 3\,\tilde{S}^{+}(np) \quad , \tag{23c}$$

$$S^{-}(\pi^{+}P) = \tilde{S}^{-}(pp) - \tilde{S}^{-}(np) \quad , \tag{23d}$$

$$S^{\pm}(K^{+}P) = 2\,\tilde{S}^{\pm}(pp) + \tilde{S}^{\pm}(np) \pm 3\,\tilde{S}^{\pm}(\lambda p) \quad , \tag{23e}$$

$$S^{\pm}(K^{+}N) = \tilde{S}^{\pm}(pp) + 2\,\tilde{S}^{\pm}(np) \pm 3\,\tilde{S}^{\pm}(\lambda p) \quad , \tag{23f}$$

with the definitions

$$\tilde{S}^{\pm}(AB) = S^{\pm}(AB) - \sigma_A(\bar{A}B) \quad ,$$

$$S^{\pm}(AB) = \sigma_T(\bar{A}B) \pm \sigma_T(AB) \quad . \tag{24}$$

In (24) A and B may refer to hadrons as well as to quarks. In all equations (23) - (24) the upper or the lower signs have to be taken simultaneously. For meson-baryon scattering $\tilde{S}^{\pm}$ and $S^{\pm}$ are of course identical. In writing down (23) we have used isospin and charge conjugation invariance, and the optical theorem (7). Equations (23) imply four relations between the quantities $\tilde{S}^{\pm}$, namely [8]

$$\tilde{S}^{+}(PP) + \tilde{S}^{+}(PN) = 3\, S^{+}(\pi^{+}P) \quad , \tag{25a}$$

$$\tilde{S}^{+}(PP) - \tilde{S}^{+}(PN) = S^{+}(K^{+}P) - S^{+}(K^{+}N) \quad , \tag{25b}$$

$$S^{-}(K^{+}P) - S^{-}(K^{+}N) = S^{-}(\pi^{+}P) \quad , \tag{25c}$$

$$\tilde{S}^{-}(PP) - \tilde{S}^{-}(PN) = S^{-}(\pi^{+}P) \quad . \tag{25d}$$

Relation (25c) is the "good" Johnson-Treiman relation [2], which is known to be in excellent agreement with the facts. As to relations (25a,b,d) a detailed comparison with experiment has to await better data on nucleon-antinucleon annihilation at high energy. This is in particular true for (25b,d) which involve cross-section differences which are of the order of the error in (22) so that a significant test is not possible at present; these relations are however fully compatible with the data. To compare relation (25a) with the data we make use of (22), which can be written as $\tilde{S}^{-}(PP) \simeq 0$. Using this in (23a,b,c) we get $\tilde{S}^{-}(PN) \simeq - S^{-}(\pi^{+}P)$, so that $\tilde{S}^{+}(PP) \simeq 2\sigma_T(PP)$ and $\tilde{S}^{+}(PN) \simeq 2\sigma_T(PN) - S^{-}(\pi^{+}P)$. Remembering that we have to compare nucleon-nucleon and meson-nucleon cross sections at laboratory momenta in the ratio $\sim \frac{3}{2}$, we find for the left and right hand sides of (25a) respectively

$$157 \pm 3\text{mb}, \quad 157.8 \pm 0.6 \text{ mb} \quad \text{at} \quad p_L^{\pi} = 8 \text{ GeV/c}$$
$$p_L^{P} = 12 \text{ GeV/c} \quad ,$$

$$154 \pm 3\text{mb}, \quad 150.0 \pm 0.6 \text{ mb} \quad \text{at} \quad p_L^{\pi} = 12 \text{ GeV/c}$$
$$p_L^{P} = 18 \text{ GeV/c} \quad ,$$

where we have used the data of Ref.[17]. Note that the agreement would have been bad if annihilation contributions would have been included on the left-hand side of (25a), increasing the latter by 15-20%. Many other relations between cross sections follow from the additivity assumption (plus only isospin and C invariance), which, however, involve one or more unmeasured cross sections [2], [3]. We will not write these down here.

Inserting the data on the left of (23) we can find numbers for the quark "cross sections". At 10 GeV/c we get, taking $\tilde{S}^{-}(PP) = 0$ and using the data of Ref.[17],

$$\tilde{S}^{-}(pp) \simeq 0.9 \text{ mb}, \ \tilde{S}^{-}(np) \simeq -1.1 \text{ mb}, \ \tilde{S}^{-}(\lambda p) \simeq -1.5 \text{ mb} \ ,$$

$$\tilde{S}^{+}(pp) \simeq 9.5 \text{ mb}, \ \tilde{S}^{+}(np) \simeq 7.7 \text{ mb}, \ \tilde{S}^{+}(\lambda p) \simeq 4.4 \text{ mb} \ .$$

(26)

From this we find for the non-annihilation parts of the quark "cross sections", $\sigma(qq')$,

$$\sigma(\bar{p}p) \simeq 5.2 \text{ mb}, \ \ \sigma(\bar{p}n) \simeq 3.3 \text{ mb}, \ \sigma(\bar{\lambda}p) \simeq 1.5 \text{ mb} \ ,$$

$$\sigma(pp) \simeq 4.3 \text{ mb}, \ \ \sigma(pn) \simeq 4.4 \text{ mb}, \ \sigma(\lambda p) \simeq 3 \ \text{ mb} \ .$$

(27)

At this stage we have to make a remark about the signs of $\tilde{S}^-(qq')$. It is reasonable to assume that the considerations of Section I.2. can also be applied to quark scattering. Remembering that by definition the $\tilde{S}^-(qq')$ do not involve quark-antiquark annihilation contributions, so that they can be only non-zero because there is a difference in the number of ways (integral) charge and/or hypercharge can be exchanged in $q + q'$ and $\bar{q} + q'$ collisions, one predicts, using (20),

$$\tilde{S}^-(pp) > 0, \quad \tilde{S}^-(np) < 0, \quad \tilde{S}^-(\lambda p) < 0. \tag{28}$$

For instance, the first inequality arises because charge can be exchanged in more ways between p and $\bar{p}$ than between p and p. We see that these predictions are satisfied by (26); it can be verified, assuming (22), that this is true for the whole momentum range $6 \leq p_L \leq 18$ GeV/c. This is consistent with the fact that quark-antiquark annihilation contributions (Fig. 2a) are not included on the right of (23), in the sense that if one would have found from (23) $\tilde{S}^-(\lambda p)$ or $\tilde{S}^-(np)$ positive this would have been an indication that in (23) the quantities $S^\pm(qq')$ which include $q\bar{q}$ annihilation, rather than $\tilde{S}^\pm(qq')$, should have occurred. Of course, the results given by (26) do not by themselves necessarily require the absence of $q\bar{q}'$ annihilation on the right-hand side of (23), but indicate that annihilation could certainly not dominate, at least for np and λp; one should always have $\sigma_A(\bar{n}p)<|\tilde{S}^-(np)|$ and $\sigma_A(\bar{\lambda}p)<|\tilde{S}^-(\lambda p)|$.

From (23) we can obtain further relations between measured cross sections if we make some extra assumptions concerning the $\tilde{S}^\pm(qq')$. In the first place, assuming in (23) the condition

$$\tilde{S}^+(pp) + \tilde{S}^+(np) = 4\,\tilde{S}^+(\lambda p) \quad , \tag{29}$$

we obtain [8]

$$3\, S^+(\pi^+P) = 2\left[S^+(K^+P) + S^+(K^+N)\right] \quad . \tag{30}$$

This relation is remarkably successful in the whole range $6 \lesssim p_L \lesssim 18$ GeV/c for instance, at 10 GeV/c we have left and right 154 mb and 156 mb respectively. We thus have the empirical fact that the quark cross sections satisfy the condition (29); its dynamical meaning, however, is unclear. It is interesting to write relations (25a) and (30) in slightly different form by introducing the average cross sections $\tilde{S}_P$, S_π, S_K, $\tilde{S}_p$, $\tilde{S}_\lambda$, defined as

$$\tilde{S}_P = \frac{1}{4}[\tilde{S}^+(PP) + \tilde{S}^+(PN)]$$

$$S_\pi = \frac{1}{2}\, S^+(\pi^+P)$$

$$S_K = \frac{1}{4}[S^+(K^+P) + S^+(K^+N)]$$

$$\tilde{S}_p = \frac{1}{4}[\tilde{S}^+(pp) + \tilde{S}^+(pn)]$$

$$\tilde{S}_\lambda = \frac{1}{2}\, \tilde{S}^+(\lambda p) \quad . \tag{31}$$

We can then write (25a), (29) and (30) as

$$\tilde{S}_P : S_\pi : S_K = 6 : 4 : 3 \quad , \tag{32}$$

$$\tilde{S}_p : \tilde{S}_\lambda = 2 : 1 \quad , \tag{33}$$

which are excellently verified from ~ 6 GeV/c up to the highest present accelerator energies. Extrapolating to $s \to \infty$ and using Pomeranchuk's theorem, (32) becomes

$$\sigma_T^\infty(PP) : \sigma_T^\infty(\pi P) : \sigma_T^\infty(KP) = 6 : 4 : 3 \quad . \tag{34}$$

Next, SU(3) invariance, i.e.,

$$\tilde{S}^{\pm}(\lambda p) = \tilde{S}^{\pm}(np) \quad , \tag{35}$$

gives two further relations, namely

$$\frac{1}{2}\, S^{-}(K^{+}P) = S^{-}(\pi^{+}P) \tag{36}$$

$$\tilde{S}^{+}(PP) = 2\, S^{+}(\pi^{+}P) - \frac{1}{2}\, S^{+}(K^{+}P) \quad . \tag{37}$$

Comparing (35) with (26) we see that SU(3) is badly broken for the quark cross sections, reflecting the well-known fact that the Johnson-Treiman relation (36) is not as good as (25c)*. Relation (37) agrees very well with the facts, the reason being that it is much less sensitive to symmetry breaking than (36). Again the agreement would have been much worse if annihilation contributions were included on the left-hand side.

Finally, we consider briefly reactions involving deuterons. Treating the deuteron (denoted by D) as a single isoscalar object and using additivity and isospin invariance one obtains the relation, analogous to (25a)

$$\tilde{S}^{+}(PD) = \frac{3}{2}\, S^{+}(\pi^{+}D) \quad , \tag{37a}$$

where the tilde on the left-hand side indicates that contributions due to annihilation of P with one of the nucleons composing the deuteron, should be subtracted from $\sigma_T(\bar{P}D)$. Because of lack of data for these contributions we cannot test (37a) directly. However, using the Glauber formula

---

* Strictly speaking SU(3) is not needed to obtain (36); what is required is the condition $\tilde{S}^{-}(\lambda p) = \tilde{S}^{-}(np)$ which is not necessarily a reflection of SU(3) symmetry [7].

$$\sigma_T(AD) = \sigma_T(AP) + \sigma_T(AN) - \frac{\langle r^{-2}\rangle}{4\pi}\,\sigma_T(AP)\sigma_T(AN) \quad ,$$

where r measures the separation of nucleons in the deuteron, and noting that for the non-annihilation parts the Glauber correction terms are to first approximation the same for A = P and A = $\bar{P}$, we have

$$\tilde{S}^-(PD) \approx \tilde{S}^-(PP) + \tilde{S}^-(PN) \approx - S^-(\pi^+P),$$

where we have used (22) and (23). Therefore $\tilde{S}^+(PD) \approx$ $\approx 2\sigma_T(PD) - \tilde{S}^-(\pi^+P)$. With this result we find, using the data of Ref. [17], at p_L^π = 10, 12, 14 GeV/c and $p_L^P = \frac{3}{2}p_L^\pi$, $\tilde{S}^+(PD)/S^+(\pi^+D)$ = 1.48, 1.50, 1.51, respectively, to be compared with the value 1.50 predicted by (37a).

We may conclude that the quark model, with additivity assumption for the non-annihilation parts of the elastic amplitudes, is in excellent agreement with the data as far as high-energy total cross sections are concerned. The model has also been successfully applied to a number of inelastic two-body reactions [27] - [28], although in most of these cases complicating features like spins and form factors, together with inaccurateness of the data, make it difficult to test additivity proper without recourse to further assumptions*. Why this simple model works so well is a mystery; a priori dynamical justifications for it are lacking. However, as will be seen below, by considering non-forward directions

---

* For inelastic reactions of the type $P\bar{P} \to B'\bar{B}$, where B and B' are baryons or baryon resonances, there is a considerable (in some cases perhaps even dominating) contribution from annihilation channels in the corresponding sum in (16), which should be excluded from additivity. This contribution is, in contrast to the case of $B'\bar{B} \to B'\bar{B}$ for t = 0 where it is $\sigma_A(B'\bar{B})$, not a measurable quantity, and the same is true for the non-annihilation part. This implies that the additivity model does not make much sense for these particular inelastic reactions, neither for the corresponding crossed reactions (see p. 155)

one may obtain in certain extreme cases a posteriori justifications for the additivity hypothesis.

We shall discuss briefly the implications of (21) for non-forward directions [3] - [5] ($t \neq 0$, but limited to small values, $|t| \lesssim 1(\mathrm{GeV}/c)^2$) in the limit $s \to \infty$. For more details see Ref. [5]. We shall assume that in this limit the quark-quark elastic scattering amplitude becomes purely imaginary (diffractive scattering) and that the total cross section becomes energy-independent (see Ch. I). Furthermore we assume that the generalized Pomeranchuk theorem (13) is valid for quarks and that only contributions of type c) (Section I.2) contribute at $s = \infty$. Restricting ourselves to pion-nucleon and nucleon-nucleon scattering, i.e., non-strange quarks, we can then write asymptotically (see remark following (21))

$$T'_{AB}(s,t) \overset{s\to\infty}{=} i\, g(t) [\sum_i f_i^A(t)] [\sum_j f_j^B(t)] \quad , \tag{38}$$

with

$$T'_{ij}(s,t) \overset{s\to\infty}{=} i\, g_{ij}(t) \qquad (g_{ij} \text{ real}) \tag{39}$$

$$g_{\bar{p}p} = g_{\bar{n}n} = g_{pp} = g_{nn} = g_{pn} = g_{np} = g_{\bar{p}n} = g_{p\bar{n}} = g \ . \tag{40}$$

In (40) we have used isospin and charge conjugation invariance. Equation (38) implies immediately the factorization theorem

$$T'_{PP}(s,t)\, T'_{\pi\pi}(s,t) = [T'_{\pi\, P}(s,t)]^2 \quad , \tag{41}$$

derived previously in the Regge pole model under the assumption of one-pole dominance [30]. If SU(3) were exact, (40) could be extended to include also the strange quarks and we would have a factorization property for all hadrons.

From (38) we can obtain in two extreme cases an understanding of why additivity can be a good approximation [3] - [5]. Case I is obtained when the entire t dependence of $\lim_{s\to\infty} T'_{AB}(s,t)$ in the region of the diffraction peak comes from the quark-quark amplitude g(t), i.e., when in that region the form factors f are approximately constant and equal to one. In that case (38) becomes

$$T'_{AB}(s,t) \overset{s\to\infty}{=} i\, g(t)\, n_A\, n_B \quad , \tag{42}$$

where n_A and n_B are the number of quarks composing A and B. This equation implies that all non-strange hadrons have asymptotic diffraction curves of the same shape, in accordance with the data (see Ch. I). Moreover, this common value of a (see (14)) for the hadrons is identical to that for quark-quark scattering. Since the size of a particle is related to the slope of its diffraction curve, this means that in this extreme case quarks have roughly the same size as hadrons. At the same time, since we have

$$\sigma_T^\infty(qq) = \frac{1}{9}\, \sigma_T^\infty(PP) \quad , \tag{43}$$

the quarks have very weak absorption, i.e., are very transparent. It is this high transparency which in the present picture may give an a posteriori justification for the additivity assumption.

Case II is the opposite of Case I. Now the t dependence of $\lim_{s\to\infty} T'_{AB}(s,t)$ in the diffraction peak is assumed to be entirely determined by the form factors in (38), i.e., g(t) is constant in t. In this case the quarks have spatial dimensions which are small compared to those of the corresponding hadrons. This small size of the quarks makes multiple scattering effects rare, so that also in this extreme picture we can understand why additivity works. A remarkable relation [4], [5] is obtained by noting that the form factors $f_i^{A,B}(t)$ should, in this limiting case of small quarks composing large hadrons, be proportional to the electromagnetic form factors $G_E^{A,B}(t)$ for hadrons A and B. Using this in (38) with $g(t) \equiv g(0)$ for small t we obtain, for the case of proton-proton scattering,

$$\frac{d\sigma_{el}(PP)}{dt} = \left(\frac{d\sigma_{el}(PP)}{dt}\right)_{t=o} [G_E^P(t)]^4 \quad , \tag{44}$$

which as shown by Fig.4 is in excellent agreement with experiment for the range $|t| \lesssim 1(\text{GeV}/c)^2$*.For πP scattering the corresponding relation cannot be tested because data on $G_E^\pi(t)$ are not available at present.On the other hand, from the experimental fact that the πP and PP diffraction curves seem to have similar shapes at very high energy we may conclude from relations of type (44) that the same is true for G_E^π and G_E^P. It should be mentioned that relation (44) can also be made plausible in Case I [4], [5].

2. Quarks and high-energy annihilation

In the foregoing we have excluded the annihilation part $\sigma_A(\bar{B}'B)$ of antibaryon-baryon cross sections from the additivity assumption. The question arises whether for annihilation at high energy one can assume some other simple mechanism through which it is possible to express $\bar{B}'B$ annihilation in terms of $\bar{q}q$ annihilation, and in so doing to relate the various antibaryon-baryon annihilation cross sections to each other. The simplest assumption one can propose is the factorization assumption [8], in which one writes the $\bar{B}'B$ annihilation amplitude as a product of three $\bar{q}'q$ annihilation amplitu-

* Using completely different arguments, Wu and Yang [31] derived (44) for the case of large t, where agreement with experiment seems to be less good.

des, momentum and energy being approximately conserved in each of the latter. It amounts to approximating Fig. 1 by Fig. 3. The assumption is compatible with the ex-

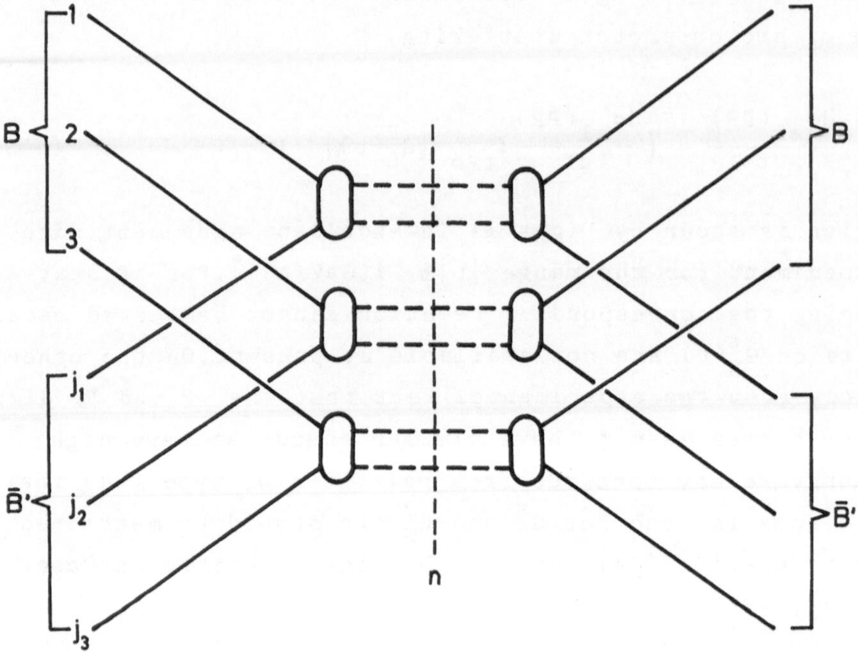

Fig. 3

perimental fact that for $\bar{P}P$ annihilation in flight the dominant contributions to $\sigma_A(\bar{P}P)$ come from states with six or more mesons (resonances counted as single mesons) in the final state [25].

The factorization assumption for $\bar{B}'B$ annihilation can be expressed in mathematical form as follows:

$$\sigma_A(\bar{B}'B) = \sum_{\{j_1 j_2 j_3\}} \prod_{i=1,2,3} Z(j_i, i) \quad , \tag{45}$$

where j_1, j_2, j_3 run over the permutations of 1, 2, 3

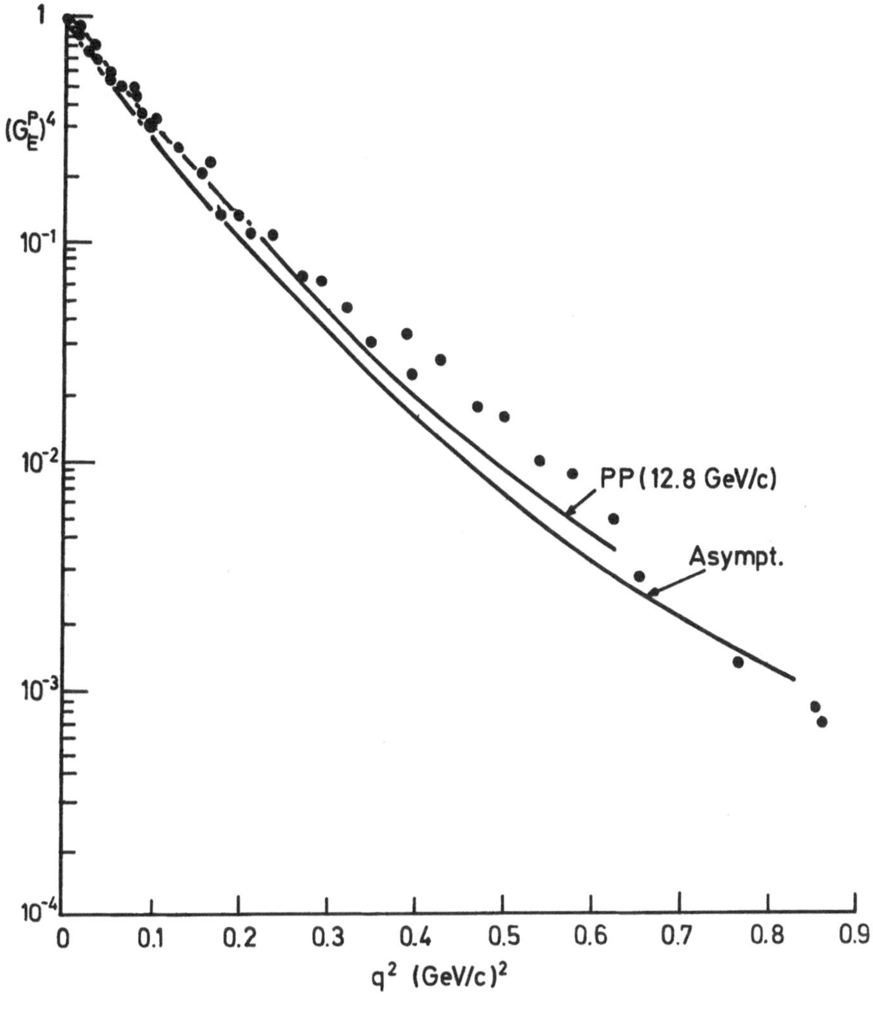

Fig. 4

Demonstration of relation (44). The full lines represent $[d\sigma_{el}(PP)/dt]/[d\sigma_{el}(PP)/dt]_{t=o}$ at 12.8 GeV/c and the asymptotic curve obtained by extrapolating PP and $\bar{P}P$ diffraction curves to infinite energy. The dots give the 4-th power of the nucleon electromagnetic form factor.

and $Z(j_i,i)$ describes the contribution of all annihilations of antiquark j_i in $\bar{B}'$ with quark i in B (see Fig. 3). Each Z, which has dimension of $(mb)^{1/3}$, is the product of an antiquark-quark annihilation cross section and a factor of dimension $(length)^{-4/3}$ determined by the wave function of the baryon as three-quark bound state. This factor is assumed to be the same for all baryons. Its explicit form is not needed for our purpose.

Using isospin and charge conjugation invariance and writing

$$Z_1 = Z(\bar{p}p), \quad Z_2 = Z(\bar{p}n), \quad Z_3 = Z(\bar{\lambda}p) \quad , \tag{46}$$

one has the following expressions for the annihilation cross sections for reactions with protons as targets:

$$\sigma_A(\bar{P}P) = 2\, Z_1^3 + 4\, Z_1 Z_2^2 \quad ,$$

$$\sigma_A(\bar{N}P) = 2\, Z_2^3 + 4\, Z_1^2 Z_2 \quad ,$$

$$\sigma_A(\bar{\Lambda}P) = \sigma_A(\bar{\Sigma}^o P) = 2\, Z_1 Z_2 Z_3 + 2\, Z_1^2 Z_3 + 2\, Z_2^2 Z_3 \quad ,$$

$$\sigma_A(\bar{\Sigma}^+ P) = 2\, Z_1^2 Z_3 + 4\, Z_1 Z_2 Z_3 \quad ,$$

$$\sigma_A(\bar{\Sigma}^- P) = 2\, Z_2^2 Z_3 + 4\, Z_1 Z_2 Z_3 \quad ,$$

$$\sigma_A(\bar{\Xi}^- P) = 2\, Z_1 Z_3^2 + 4\, Z_2 Z_3^2 \quad ,$$

$$\sigma_A(\bar{\Xi}^o P) = 2\, Z_2 Z_3^2 + 4\, Z_1 Z_3^2 \quad ,$$

$$\sigma_A(\bar{\Omega}^- P) = 6\, Z_3^3 \quad . \tag{47}$$

At present, experimental information at high-energy exists only for the first cross section of the list. This means that we cannot yet compare the factorization assumption with experiment.

It will be clear that neither the additivity assumption as used here nor the factorization assumption by themselves can produce relations for hadron cross sections that connect annihilation and non-annihilation effects. Relations of that type can only be obtained if one makes such a connection "ad hoc" on the level of the quarks. Examples of such relations are the Freund relations [6], [7], [32]

$$S^-(PP) = 5\, S^-(\pi^+P) \quad ,$$

$$S^-(PN) = 4\, S^-(\pi^+P) \quad , \tag{48}$$

and the Levinson-Wall-Lipkin relations [33]

$$S^-(PP) = 2\, S^-(K^+P) + S^-(K^+N) \quad ,$$

$$S^-(PN) = S^-(K^+P) + 2\, S^-(K^+N) \quad , \tag{49}$$

which essentially express nucleon-antinucleon annihilation in terms of meson-nucleon scattering, $S^-(PP)$ and $S^-(PN)$ being dominated by annihilation contributions. It is easy to see, using (22), (23) and (47), that (49) and (48) imply the following relations between quark annihilation and non-annihilation cross sections: from (48)

$$Z_1^3 = Z_2^3 = -\frac{3}{2}\,\tilde{S}^-(np) \quad , \tag{50}$$

and from (49)

$$Z_1^3 = Z_2^3 = -\frac{3}{2}\,\tilde{S}^-(\lambda p) \quad . \tag{51}$$

Remember that both right-hand sides are positive (see (28)). The dynamical meaning of conditions (50) and (51) is unclear; they should be regarded as expressing within the model the empirical facts that (48) is rather well and (49) very well satisfied by the data*. As such their status is analogous to, for instance, (29).

The equality $Z_1 = Z_2$ implied by (50) or (51), used in (47), leads to

$$\frac{\sigma_A(o)}{\sigma_A(1)} = \frac{\sigma_A(1)}{\sigma_A(2)} = \frac{\sigma_A(2)}{\sigma_A(3)} \tag{52}$$

where $\sigma_A(S) = \sigma_A(\bar{B}P)$ for strangeness S of $\bar{B}$, σ_A being independent of the charge and isospin of $\bar{B}$.

III. Regge Pole Model and Current Algebra

In this last part we shall discuss a model, proposed recently by Cabibbo, Horwitz and Ne'eman (CHN) [9], which essentially consists of the Regge pole model endowed with an algebraic structure for the residue functions, or more precisely for the vertex strength functions. The original motivation for this model was to obtain an algebraic foundation for the successful high energy results produced by the quark model (Ch. II). Up to now the model has been formulated only for elastic scattering in the forward direction with absence of spins.

The model is based on the following assumptions:

1. The Regge pole model for elastic scattering, i.e., the amplitude $T_{AB}(s,t)$ for reaction (1) at high energy

---

* In the original form of the quark model in which annihilation was supposed to be included in the additivity assumption, (48) and (49) were obtained by putting $S^-(np) = 0$ and $S^-(\lambda p) = 0$ respectively [6], [7], [33]. By writing them as $\sigma_A(\bar{n}p) = -\tilde{S}^-(np)$ and $\sigma_A(\bar{\lambda}p) = -\tilde{S}^-(\lambda p)$ it is clear that, analogously to (50) and (51), the nature of these conditions is to express for quark scattering annihilation contributions in terms of non-annihilation contributions.

can be written in the form (3) with

$$c_j^{AB}(t) = -\beta_j^{AB}(t)\ \xi_j^{\pm}(t)\ \frac{\Gamma(\alpha_j(t)+\frac{3}{2})}{\Gamma(\alpha_j(t)+1)} \quad . \tag{53}$$

In here $\beta_j^{AB}(t)$ is the residue function associated with the Regge pole $\alpha_j(t)$. Using the factorization theorem [30] we can express this function as a product of two vertex strength functions, as follows:

$$\beta_j^{AB}(t) = \gamma_j^{A}(t)\ \gamma_j^{B}(t) \tag{54}$$

$\xi_j^{\pm}(t)$ is the signature factor, given by

$$\xi_j^{\pm}(t) = \frac{1 \pm e^{-i\pi\alpha_j(t)}}{\sin \pi\alpha_j(t)} \quad , \tag{55}$$

where the plus (minus) sign has to be taken if the physical states on the Regge trajectory $\alpha_j(t)$ have even (odd) angular momentum.

2. Two unitary nonets of Regge trajectories dominate high-energy scattering, one, $s_j(j=0,...,8)$, of even signature having the 2^+ mesons (A_2, K^{**}, f_o, f_o') as physical particles, and one, $v_j(j = 0,...,8)$, of odd signature corresponding to the nonet of 1^- mesons (ρ, K^*, ω, ϕ). The Pomeranchuk trajectory, describing the asymptotic behaviour of the amplitude, is assumed to be included in the even signature nonet.

3. The functions $\gamma_j^{A,B}$ occurring in (54) are for t=0 given by the matrix elements of operators S_j and V_j belonging to a U(12) algebra,

$$\left.\begin{aligned} \delta^3(\vec{p}_A - \vec{p}_A')\gamma_{s_j}^{A}(o) &= \langle A|S_j|A\rangle \qquad (56) \\ \delta^3(\vec{p}_A - \vec{p}_A')\gamma_{v_j}^{A}(o) &= \langle A|V_j|A\rangle \qquad (57) \end{aligned}\right\} \quad j = 0,...,8$$

with

$$S_j = \int D(\beta\lambda_j, \vec{x}, o)\, d^3x \quad , \tag{58}$$

$$V_j = \int D(\ \lambda_j, \vec{x}, o)\, d^3x \quad , \tag{59}$$

the integrals being carried out in the rest frame of the incident particle. The D's are current densities, which in a quark representation have the form

$$D(M_j, \vec{x}, o) = \frac{1}{2}\, q^+ M_j q \quad , \tag{60}$$

where q and q^+ are the quark fields. The densities associated with S_j and V_j are of scalar and vector type respectively. The V_j's have the same algebraic properties as the unitary spin generators. Together with the S_j they form a U(3) × U(3) algebra isomorphic to the $[U(3) \times U(3)]_\beta$ contained in the $[U(6) \times U(6)]_\beta$ of Dashen and Gell-Mann [34]. They satisfy

$$[S_i, S_j] = i\, f_{ijk}\, V_k \quad ,$$

$$[V_i, S_j] = i\, f_{ijk}\, S_k \tag{61}$$

with the f_{ijk} (just as the unitary matrices λ_i) defined in Ref. [35]. These commutation relations fix the relative scales of γ_{S_j} and γ_{V_j}. For elastic scattering, we need only j = 0, 3, 8 in (56) and (57), j = 0 referring to the unitary singlet, j = 3, 8 to the centre elements of the octet. Hence each elastic amplitude $T_{AB}(s,t)$ is, in general, a sum of six terms, three arising from the even signature nonet, coupled to the quantities S_j, in which j = 0 corresponds to the Pomeranchuk trajectory, and three from the odd signature nonet, coupled to the quantities V_j.

On the basis of the above assumptions the following

expressions for the total cross sections are obtained:

$$
\begin{aligned}
S^{\pm}(PP) &= 12\, t_o^{\pm} + 6\, t_8^{\pm} + 2\, t_3^{\pm} \quad, \\
S^{\pm}(PN) &= 12\, t_o^{\pm} + 6\, t_8^{\pm} - 2\, t_3^{\pm} \quad, \\
S^{+}(\pi^+P) &= 8\, t_o^{+} + 4\, t_8^{+} \quad, \\
S^{-}(\pi^+P) &= 4\, t_3^{-} \quad, \\
S^{+}(K^+P) &= 8\, t_o^{+} - 2\, t_8^{+} + 2\, t_3^{+} \quad, \\
S^{-}(K^+P) &= 6\, t_8^{-} + 2\, t_3^{-} \quad, \\
S^{+}(K^+N) &= 8\, t_o^{+} - 2\, t_8^{+} - 2\, t_3^{+} \quad, \\
S^{-}(K^+N) &= 6\, t_8^{-} - 2\, t_3^{-} \quad.
\end{aligned}
\tag{62}
$$

In here the quantities $t_j^{\pm}$ are given by

$$
t_j^{\pm} = \frac{\Gamma(\alpha_j^{\pm} + \frac{3}{2})}{\Gamma(\alpha_j^{\pm} + 1)} \left(\frac{\nu}{\nu_j^{\pm}}\right)^{\alpha_j^{\pm}(o)} \quad, \tag{63}
$$

with $\nu = s_o = s - m_A^2 - m_B^2$ and where we have explicitly taken out scale factors $\nu_j^{\pm}$. The latter are unknown parameters. The superscript $\pm$ refers to the signature. In calculating the matrix elements in (56) and (57) the $[U(6) \times U(6)]_\beta$ assignments of $(56,1)^+$ and $(6,6^*)^-$ for baryons and mesons are used, which in the quark model correspond to the usual quark composition of the hadrons. The normalization $tr\lambda_i^2 = 2$ is used. The signs in (62) are determined by the signature factors (55) and by the requirement that at the physical poles the exchange of 2^+ and 1^- particles gives rise to forces between equal particles which are respectively attractive and repulsive. The non-trivial assumption is then made that the extrapolation to $t = 0$ is not accompanied by changes in sign. If such a change in sign occurs between $t = 0$ and the physical pole the corresponding $t_j^{\pm}$ in (62) has to be replaced by $- t_j^{\pm}$(see remark 4 below).

No F/D and octet-singlet mixing is as yet taken into account (see below); (62) corresponds to pure F coupling for the vector trajectories, while the coupling of the scalar trajectories to baryons and mesons is pure F and D respectively.

From (62) the following four relations between cross sections are obtained after eliminating $t_j^{\pm}$:

$$S^+(PP) + S^+(PN) = 3\,S^+(\pi^+P) \quad , \tag{64a}$$

$$S^+(PP) - S^+(PN) = S^+(K^+P) - S^+(K^+N) \quad , \tag{64b}$$

$$S^-(K^+P) - S^-(K^+N) = S^-(\pi^+P) \quad , \tag{64c}$$

$$S^-(PP) - S^-(PN) = S^-(\pi^+P) \quad . \tag{64d}$$

These relations have to be compared with the quark model results (25). They only differ in the fact that, in contrast to (25), they include annihilation contributions. As mentioned in Ch. II, this affects mainly (64a), which around 10 GeV/c has a discrepancy with experiment of the order of $\sim$ 20 %. Of course, asymptotically (25a) and (64a) become identical and predict $\sigma_T(PP)/\sigma_T(\pi P) = \frac{3}{2}$. As to relations (64b and d), due to the large experimental uncertainties in $\sigma_T(\bar{P}N)$ similar remarks apply to them as made with respect to (25b and d). We come back to relations (64) at the end of this Chapter.

With the signs in (62) fixed by the above-mentioned assumption a number of inequalities follows; examples are

$$S^-(AP) > 0, \qquad \text{for } A = K^+, \pi^+, P,$$

$$S^+(AP) > S^+(AN) \quad , \qquad \text{for} \quad A = K^+, P \quad ,$$

$$S^-(PP) > S^-(PN) \quad ,$$

$$\sigma_T(\bar{P}P) > \sigma_T(PN) \quad , \qquad \sigma_T(\bar{P}P) > \sigma_T(\bar{P}N) \quad , \tag{65}$$

which are all in accordance with the data above 6 GeV/c.

Other relations are obtained from (62) by assuming equalities among the $t_j^{\pm}$. For instance, U(3) symmetry for the vector contributions, i.e., $t_3^- = t_8^- = t_o^-$, is equivalent to the Freund relations (48); the equality $t_o^- = t_8^-$ implies the Levinson-Wall-Lipkin relations (49). Exchange degeneracy [36] for the I = 1 exchanges, i.e., $t_3^- = t_3^+$, gives

$$\sigma_T(K^+P) = \sigma_T(K^+N) \quad ,$$

which seems to agree very well with the facts [17]*. SU(3) for the vector and scalar trajectories gives respectively the Johnson-Treiman relation (36), and

$$S^+(PP) = 2\, S^+(\pi^+P) - \frac{1}{2}\, S^+(K^+P) \quad . \tag{66}$$

The latter relation is the counterpart of (37); it is, however, less well satisfied by the data.

The model also predicts another interesting property [10]. A good fit to the entire data on total cross sections can only be obtained within the model if one assumes all σ_T to decrease at very high energy as

$$\sigma_T \sim \nu^{-(0.075 \pm 0.008)} \quad , \tag{67}$$

corresponding to $\alpha_o^+(o) = 0.925 \pm 0.008$, where $\alpha_o^+(o)$ is the intercept of the Pomeranchuk trajectory. This can

* In the quark model this equality is implied by the equality of σ(pp) and σ(np) (see (27) and (74) below).

best be seen by considering the average cross sections $S_P = \frac{1}{4}[S^+(PP) + S^+(PN)]$ and S_π and S_K , defined in (31), which in terms of t_j^+ become

$$S_P = 6\, t_o^+ + 3\, t_8^+ \quad , \tag{68a}$$

$$S_\pi = 4\, t_o^+ + 2\, t_8^+ \quad , \tag{68b}$$

$$S_K = 4\, t_o^+ - t_8^+ \quad , \tag{68c}$$

with

$$t_j^+(\nu) = t_j^+(1)\,\nu^{\alpha_j^+(o)-1} \tag{69}$$

If one assumes $\alpha_o^+(o) = 1$, i.e., $t_o^+ = \text{const.} \neq 0$, Eqs. (68) predict that S_P and S_π approach their asymptotic limits $6\, t_o^+$ and $4\, t_o^+$ from above, but S_K approaches its limit $4\, t_o^+$ from below. This runs counter to the experimental fact that for $p_L \gtrsim 5$ GeV/c all hadron total cross sections are known to decrease with increasing energy. Hence with $\alpha_o^+(o) = 1$ Eqs. (68) cannot fit the data. In fact, since from (68b and c) one has

$$\frac{1}{12}(S_\pi + 2\, S_K) = t_o^+ \quad , \tag{70}$$

a straight line fit to the experimental values for the left-hand side on a logarithmic plot (Fig. 5) gives $\alpha_o^+(o) \approx 0.93$. To fit the entire data, including the nucleon-nucleon data, it is necessary to allow for F/D admixture at the scalar trajectory nucleon vertex together with some octet-singlet mixing. The reason is that Eqs. (68) predict $S_P/S_\pi = \frac{3}{2}$, i.e., relation (64a), which, as mentioned earlier, is in disagreement with experiment (remember that S_P includes annihilation). The modified equations become

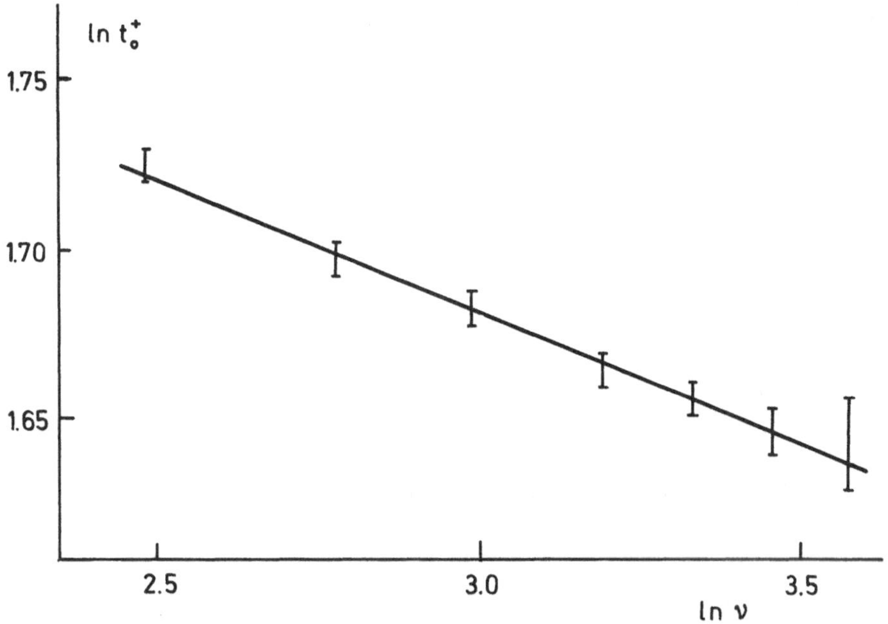

Fig. 5.

Fit of the form $t_o^+(1)\nu^{\alpha_o^+(o)-1}$ to the left-hand side of (70). The experimental points are taken from Ref. [17].

$$S_P = (\alpha\sqrt{6} + \beta\lambda\sqrt{3})^2\, t_o^+ + (\beta\sqrt{6} - \alpha\lambda\sqrt{3})^2\, t_8^+ \quad ,$$

$$S_\pi = 2(\alpha\sqrt{2} + \beta)(\alpha\sqrt{2} + \beta\lambda)t_o^+ + 2(\beta\sqrt{2} - \alpha)(\beta\sqrt{2}-\alpha\lambda)t_8^+ \; ,$$

$$S_K = (2\alpha\sqrt{2} - \beta)(\alpha\sqrt{2} + \beta\lambda)t_o^+ + 2(\beta\sqrt{2} + \alpha)(\beta\sqrt{2}-\alpha\lambda)t_8^+ \; , \qquad (71)$$

with $\lambda = F - \frac{1}{3} D$, $\alpha = \cos\phi$, $\beta = \sin\phi$, ϕ being the (t_o^+, t_8^+) mixing angle. An excellent fit to the data of Ref. [17] is obtained with the following parameters

$$t_o^+(1) = 7.43 \pm 0.56 \quad , \quad t_8^+(1) = 2.08 \pm 0.26$$

$$\alpha_o^+(o) = 0.925 \pm 0.008 \quad , \quad \alpha_8^+(o) = 0.76 \pm 0.04$$

$$\tan\phi = -0.066 \pm 0.048 \quad , \quad \lambda = 2.44 \pm 0.09 \quad . \quad (72)$$

This result remains essentially unchanged if we put $\phi = 0$. In that case we have from (71) in the limit $\nu \to \infty$, $S_A \to 0$ as $\nu^{-0.075}$ ($A = P, \pi, K$) and

$$\frac{S_P}{S_\pi} \to \frac{3}{2} \quad , \quad \frac{S_K}{S_\pi} \to 1 \quad . \quad (73)$$

With the small mixing angle given by (72) the latter values change by a few percent. Compare (73) with (34) and remember that the latter result is obtained from an extrapolation of the <u>empirical</u> relation (30), i.e., the quark model does not necessarily contradict (73) (see also remark 3 below). It should be remarked that also with Eqs. (71) no acceptable fit to the data can be obtained with $\alpha_o^+(o) = 1$.

We end with a few remarks.

1. It is possible to escape the vanishing of total cross sections by adding to the two nonets of trajectories introduced in assumption 2 above a second even signature singlet trajectory, which could then play the role of the Pomeranchuk trajectory and have intercept going through one. Good fits to the data, based on such a picture, are known to exist [37], [38]. However, such a possibility would mean an extension of the algebraic structure defined in assumption 3 and would therefore be outside the scope of the present model *. Without altering the basic algebraic structure of the model one

is led to (67). Needless to say, the above considerations do not prove the vanishing of cross sections, no more than the considerations in Ref. [37], [38] prove their constancy, but show that such a possibility, fully compatible with the existing experimental situation including cosmic ray data, exists, and that it can be linked with an economical and appealing scheme.

2. At very high energy the t_o^+ contributions will dominate each cross section, so that in a $(\log \sigma, \log \nu)$ plot the latter approach a straight line asymptotically. Certain cross sections, such as $\sigma_T(K^+N)$ and $\sigma_T(K^+P)$, will approach this asymptote from below. Because of this, they may appear to be nearly constant throughout a large energy region, as is actually the case above 6 GeV/c for the just mentioned cross sections. Finally, from the fact that all elastic amplitudes behave asymptotically as

$$T(\nu,t) \underset{\nu\to\infty}{\sim} \xi^+(t)\nu^{\alpha_o^+(t)} \quad ,$$

the value for $\alpha_o^+(o)$ given by (72) gives a non-vanishing ratio of real to imaginary part $X(\nu,o)$ at $t = 0$. One finds for all hadrons $X(\nu,0) \overset{\nu\to\infty}{=} - 0.12$. This value is compatible with existing experimental results [10],[18].

3. It is easy to verify that by forming the following combinations of quark cross sections:

---

* An additional singlet, being outside the algebra of $[U(6)\times U(6)]_\beta$, would add to the right of (62) extra terms $\Gamma_N, \Gamma_\pi, \Gamma_K$, for $S^+(P\mathcal{N}), S^+(\pi^+P), S^+(K^+\mathcal{N})$, respectively ($\mathcal{N}$ = nucleon), which would be arbitrary parameters. The model would therefore lose much of its predictive power; for instance the asymptotic prediction $\sigma(PP)/\sigma(\pi P) = \frac{3}{2}$ would not be obtained.

$$t_o^\pm = \frac{1}{4}\,[\tilde{S}^\pm(pp) + \tilde{S}^\pm(np) + \tilde{S}^\pm(\lambda p)] \quad ,$$

$$t_8^\pm = \frac{1}{4}\,[\tilde{S}^\pm(pp) + \tilde{S}^\pm(np) - 2\,\tilde{S}^\pm(\lambda p)] \quad ,$$

$$t_3^\pm = \frac{1}{4}\,[\tilde{S}^\pm(pp) - \tilde{S}^\pm(np)] \quad , \qquad (74)$$

and denoting them by $t_j^\pm$ as indicated above, one can formally put (23) into the form (62) and vice versa (see also Ref. [39]), replacing in the latter set of equations $S^\pm(\mathcal{N}\mathcal{N}')$ by $\tilde{S}^\pm(\mathcal{N}\mathcal{N}')$. In the CHN model, each $t_j^\pm$ is associated with a single Regge pole in the t channel of (1). In the quark model each combination in (74) may be associated with more than one Regge pole, or more general, may have more than one term in an expansion in powers of ν. Using (74) in (23) we can write in the quark model, analogous to (68),

$$\tilde{S}_P = 6\,t_o^+ + 3\,t_8^+ \quad ,$$

$$S_\pi = 4\,t_o^+ + 2\,t_8^+ \quad ,$$

$$S_K = 4\,t_o^+ - t_8^+ \quad . \qquad (75)$$

In general the simplest way to parametrize t_o^+ and t_8^+ is as follows: $t_o^+ = x_1 + x_2\,\nu^{x_3}$ and $t_8^+ = x_4 + x_5\,\nu^{x_6}$. The case of vanishing cross sections discussed above and in Ref. [10] corresponds to putting $x_1 = x_4 = 0$ and fitting the remaining parameters. The situation in which $S_\pi \neq S_K \neq 0$ for $s \to \infty$ (as considered in Refs. [37], [38]) requires all $x_i (i=1,\dots,6)$ to be non-zero.

4. The last remark is of speculative nature. We have seen in the foregoing Chapter that in the quark model different mechanisms control the annihilation and non-annihilation parts of the total cross sections. This rests on the fact that annihilation effects correspond to baryon exchange, while all non-annihilation processes

are of different type, corresponding to exchange of vanishing baryon number. One might also say that the latter processes are peripheral and the annihilation processes non-peripheral or central. This difference in nature of the two types of processes may require different descriptions for them, not only in the quark model, but also in other dynamical models [8]. This may be true in particular for the Regge pole model [40]. In fact, from the observation that additivity of quark amplitudes corresponds to t channel exchanges of singlet and octet type, one might speculate about the possibility that the Regge pole model in which only singlet and octet trajectories are exchanged, like the CHN model, describes only the non-annihilation part of the total cross sections. If this would be true, the relations obtained in the quark model and the CHN model would become identical (in (64) and (66) we have to replace $S^{\pm}(\mathcal{N}\mathcal{N}')$ by $\tilde{S}^{\pm}(\mathcal{N}\mathcal{N}')$, $\mathcal{N}$ = nucleon (see (24))). Since these relations are in very good agreement with experiment we would no longer need the D/F admixture in the scalar trajectory-nucleon vertex (see (72)); indeed, this parameter was introduced precisely to account for the disagreement of (64a) and (66) with the facts. The conclusion (67) would obviously remain unshaken. The cross section differences would present a difficulty. It is clear from (62) that in order to satisfy (22) one would need

$$6\ t_0^- + 3\ t_8^- + t_3^- \simeq 0 \quad . \tag{76}$$

Since from the $K^{\pm}\mathcal{N}$ and $\pi^{\pm}P$ data one has $t_3^- > 0$ and $t_8^- > 0$, (76) would imply $t_0^- < 0$, i.e., the residue function of the v_0 trajectory would have to change sign between t = 0 and the t value corresponding to the physical pole. This is perhaps not nice, but there is no compelling reason against such a sign change. In fact, there are for instance good reasons to believe that the residue function of the ρ trajectory changes sign at $t \simeq -0.2(GeV/c)^2$ [41]. It is easy to verify that t_o^-, t_8^- mixing would not alter this conclusion.

References

1. E. M. Levin and L. L. Frankfurt, Zhur.Eksp.i Teoret. Fiz. Pisma v Redak. 2, 105 (1965) - English translation: Soviet Phys., JETP Letters 2, 65 (1965)
2. H. J. Lipkin and F. Scheck, Phys. Rev. Lett. 16, 71 (1966).
3. J. J. J. Kokkedee and L. Van Hove, Nuovo Cim. 42, 711 (1966).
4. L. Van Hove, Proceedings of the Stony Brook Conference on High Energy Two-Body Reactions (April 1966).
5. L. Van Hove, Lectures at 1966 Scottish Universities Summer School, CERN Preprint TH. 676 (June 1966).
6. H. J. Lipkin, Phys. Rev. Lett. 16, 1015 (1966).
7. J. J. J. Kokkedee, Phys. Lett. 22, 88 (1966).
8. J. J. J. Kokkedee and L. Van Hove, Nuclear Phys. B1, 169 (1967).
9. N. Cabibbo, L. Horwitz and Y. Ne'eman, Phys. Lett. 22, 336 (1966).
10. N. Cabibbo, J. J. J. Kokkedee, L. Horwitz and Y. Ne'eman, Nuovo Cim. 45, 275 (1966).
11. L. Van Hove, Revs. Modern Phys. 36, 655 (1964).
12. L. Van Hove, CERN Lectures, Yellow Report 65-22 (1965).
13. L. Van Hove, High Energy Physics and Elementary Particles, International Atomic Energy Agency, Vienna, p. 179 (1965).
14. A. Białas and E. Białas, Nuovo Cim. 37, 1686 (1965).
15. I. Pomeranchuk, Zhur. Eksp. i. Teoret. Fiz. 34, 725 (1958) - English Translation: Soviet Phys., JETP 34, 499 (1958).

16. For a review of the experimental situation, see S. J. Lindenbaum, Proceedings Oxford Conference (1965).
17. W. Galbraith, et al., Phys. Rev. 138, B913 (1965).
18. K. J. Foley, et al., Phys. Rev. Lett. 14, 74, 862 (1965); G. Belletini, et al., Phys. Lett. 14, 164 (1965); 19, 341, 705 (1965); L. Kirillova, et al., Soviet Journal of Nuclear Physics 1, 379 (1965).
19. K. J. Foley, et al., Bull. Am. Phys. Soc. 12, 103 (1967).
20. L. Van Hove, Phys. Lett. 5, 252 (1963).
21. O. Czyzewski, et al., Phys. Lett. 15, 88 (1965).
22. J. J. J. Kokkedee, Nuovo Cim. 43, 919 (1966).
23. M. Gell-Mann, Phys. Lett. 8, 214 (1964); G. Zweig, CERN Preprints TH. 401, 402 (1964), unpublished.
24. For a discussion which includes spins, see C. Itzykson and M. Jacob, Saclay Preprint (July 1966); see also Refs. [27], [29].
25. K. Böckmann, et al., Nuovo Cim. 42A, 954 (1966).
26. L. I. Schiff, Phys. Rev. Lett. 17, 612, 714 (1966).
27. G. Alexander, H. J. Lipkin and F. Scheck, Phys. Rev. Lett. 17, 412 (1966).
28. J. L. Friar and J. S. Trefil, CERN Preprint TH.723 (1966).
29. K. Kajantie and J. S. Trefil, Phys. Lett. 24B, 106 (1967); H. Joos, Phys. Lett. 24B, 103 (1967).
30. M. Gell-Mann, Phys. Rev. Lett. 8, 263 (1962); V. N. Gribov and I. Pomeranchuk, Phys. Rev. Lett. 8, 343 (1962).
31. T. T. Wu and C. N. Yang, Phys. Rev. 137, B708 (1965).
32. P. G. O. Freund, Phys. Rev. Lett. 15, 929 (1965).
33. C. A. Levinson, N. S. Wall and H. J. Lipkin, Phys. Rev. Lett. 17, 1122 (1966).
34. R. F. Dashen and M. Gell-Mann, Phys. Lett. 17, 142, 145 (1965).
35. M. Gell-Mann, Phys. Rev. 125, 1067 (1962).

36. R. C. Arnold, Phys. Rev. Lett. 14, 657 (1965).
37. V. Barger and M. Olsson, Phys. Rev. 146, 1080 (1966).
38. P. G. O. Freund, Nuovo Cim. 46A, 563 (1966).
39. H. Yabuki, University of Tokyo Preprint (1966).
40. A. Białas and K. Zalewski, Nuovo Cim. 46A, 425 (1966).
41. G. Höhler, J. Baacke and G. Eisenbeiß, Phys. Lett. 22, 203 (1966).

S-MATRIX SINGULARITY STRUCTURE†

By

J. C. POLKINGHORNE

Department of Applied Mathematics and Theoretical Physics, University of Cambridge, England

1. Introduction

In these lectures I want to survey some of the basic ideas and results in the effort to determine the singularity structure of the S-matrix.

The S-matrix is the operator connecting initial and final states in a scattering experiment. Its principal properties are

1. Unitarity;
2. Lorentz Invariance;
3. Connectedness Structure;
4. Analyticity;
5. Crossing-Symmetry.

The first two properties are too familiar to call for comment. Property (3) proves to be of crucial importance. It expresses the short range nature of the forces and may be symbolically expressed by the decomposition of S-matrix elements into sums of connected

---

† Lecture given at the VI. Internationalen Universitätswochen f.Kernphysik,Schladming,26 February-11 March 1967.

parts together with lines corresponding to δ-functions expressing the fact that some particles may be unaffected by the interaction, e.g.

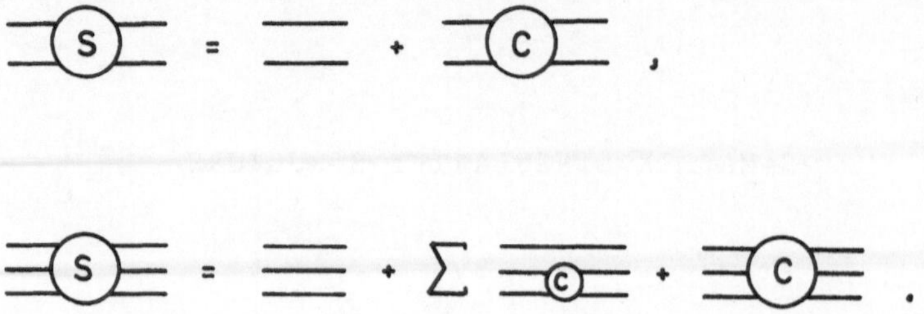

(1)

The connected parts, denoted in (1) by bubbles with C's, are then, by property (4), supposed to be analytic once an overall energy-momentum conserving δ-function has been factored out:

$$<S_{connected}> = - i(2\pi)^4 \cdot \delta(\Sigma p - \Sigma p') <A^{(+)}> ,$$

$$<S^{+}_{connected}> = i(2\pi)^4 \cdot \delta(\Sigma p - \Sigma p') <A^{(-)}> . \quad (2)$$

The principle of maximum analyticity states that the singularities of these amplitudes $A^{(\pm)}$ are "the minimum permitted by unitarity". It is the purpose of our investigation to try to give as precise a meaning as possible to this at first somewhat oracular requirement.

The principle (5) of crossing symmetry is no doubt generally familiar. It states that the same analytic function evaluated in appropriate limits and for the appropriate values of the invariants gives the scattering amplitudes corresponding to the different processes obtainable by permuting particles between initial and final states. It is intimately bound up with (4), both because only by analyticity do we obtain a global property which gives crossing meaning, and also because Olive has shown how crossing may be deduced once enough is known about the singularity structure.

The combination of 1. and 3. leads to unitary equations in the form in which we shall use them, e.g.

$(2\mu)^2 \leq (\Sigma p)^2 < (3\mu)^2$

$(3\mu)^2 \leq (\Sigma p)^2 < (4\mu)^2$

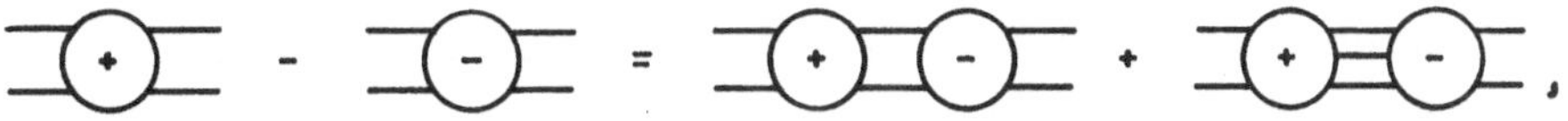

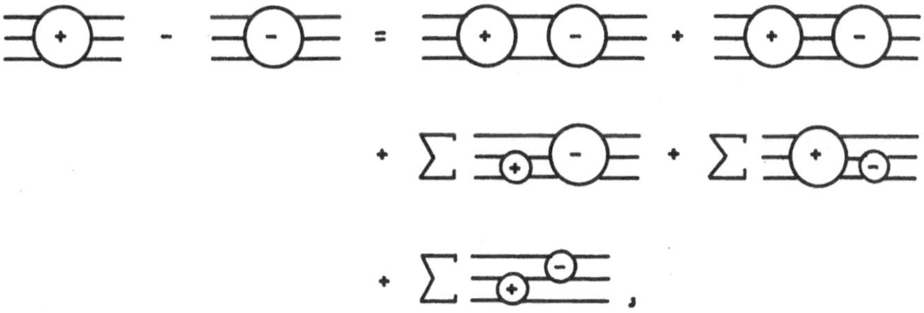

etc. (3)

In these equations we use S-matrix bubble notation:

m'  m $\longrightarrow$ $\langle m'|A^{\pm}|m\rangle$,

internal lines $\longrightarrow$ $2\pi i\delta^{(+)}(q^2-\mu^2)$,

each loop $\longrightarrow$ $\dfrac{i}{(2\pi)^4}\int d^4k$,

n identical particle intermediate state $\longrightarrow$ $\dfrac{1}{n!}$

2. Singularities of Integrals

We shall need to understand some basic ideas on the singularities of integrals. The general ideas are simple and can be illustrated by a simple single-dimensional integral

$$f(\xi) = \int_a^b dz\ F(\xi,z) \quad . \tag{4}$$

F is assumed analytic in z and ξ except for certain singularities which move in the z-plane as ξ is varied. As one follows a path in the ξ-plane the contour joining a to b is varied, using Cauchy's theorem, in such a way as to avoid the oncoming singularities of F. There will only be a singularity in ξ when these distortions are not possible and this happens in the two cases illustrated by Fig. 1:

a) end point singularity,
when a singularity of F impinges on an end point; and

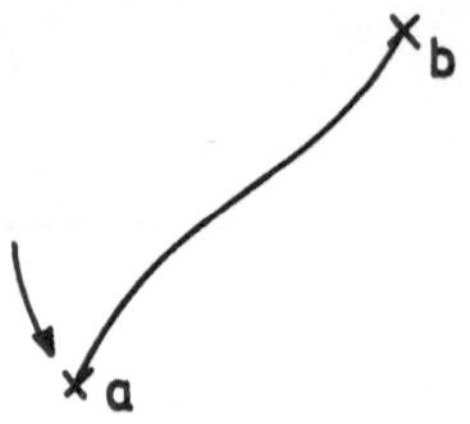

a) end point

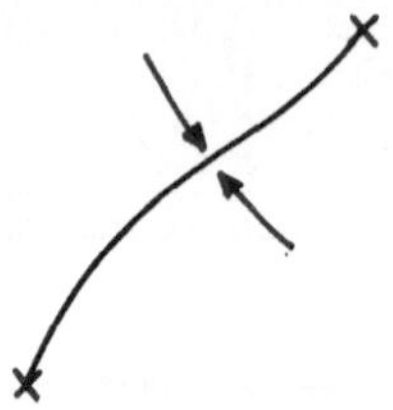

b) pinch

Fig. 1

b) pinch singularity,
when two coincident singularities of F trap the contour between them.
If the singularities of F are given by the equation

$$\sigma(\xi,z) = 0 \tag{5}$$

then necessary conditions for singularities are

$$\sigma(\xi,a) = 0 \qquad \text{or} \qquad \sigma(\xi,b) = 0 \quad , \tag{6a}$$

$$\sigma(\xi,z) = \frac{\partial\sigma}{\sigma z} = 0 \quad . \tag{6b}$$

Notice that Eq. (6b) is sufficient for a coincidence but not for a singularity since the contour must actually be trapped.

These ideas generalize readily to multiple integrals. In simple cases one just applies the foregoing argument to each successive integration. However, it will be useful to give a more sophisticated discussion. We now have a hypercontour of integration which may be distorted arbitrarily at points in its interior. Its edges are defined by certain analytic manifolds (e.g. $z_i = 0$) and can only be distorted within these manifolds.

Consider first a singularity occurring due to the hypercontour being trapped at an interior point Z by a singularity surface,

$$\Sigma = 0 \quad . \tag{7}$$

If Σ had a well defined normal at Z it would give a possible direction for distorting the hypercontour to avoid it. Therefore a necessary condition for a pinch is

$$\Sigma = 0 = \frac{\partial\Sigma}{\partial z_i} \qquad i = 1,2,\ldots \tag{8}$$

for then Σ is locally cone-like at Z and the hypercontour can be caught between the two halves of the cone. This is sufficient if the hypercontour is actually trapped.

A singularity might also arise at an interior point due to two singularity surfaces Σ_1 and Σ_2 . The contours could be trapped if Σ_1 and Σ_2 touched at Z. The condition for this would be

$$\Sigma_1 = \Sigma_2 = 0 = \lambda_1 \frac{\partial \Sigma_1}{\partial z_i} + \lambda_2 \frac{\partial \Sigma_2}{\partial z_i} \qquad i=1,2,\ldots \qquad (9)$$

In general, if F has singularity surfaces $\Sigma_1,\ldots,\Sigma_m$ they can produce pinches at an interior point of the hypercontour if

$$\lambda_j \Sigma_j = 0, \qquad j = 1,\ldots,m;$$

$$\sum_j \lambda_j \frac{\partial \Sigma_j}{\partial z_i} = 0 \quad , \quad i = 1,\ldots,n \; . \qquad (10)$$

In Eq. (10) some of the λ's may be zero in which case the corresponding surfaces do not participate in the pinch.

A simple way of seeing that Eq. (10) is correct is as follows. The distortions of the hypercontour are determined by the equations of the singularity surfaces and not by the precise nature (pole, square root, etc.) of the singularities themselves. We can therefore use, as a probe to follow the distortions, the test function

$$\frac{1}{\Sigma_1 \Sigma_2 \ldots \Sigma_m} \qquad (11)$$

which certainly has the singularity surfaces we are interested in. Now, using the Feynman trick, we can rewrite Eq. (11) as

$$(m-1)! \int_0^1 d\lambda_j \frac{\delta(\Sigma\lambda_j - 1)}{[\lambda_j \Sigma_j]^m} \quad . \qquad (12)$$

We now have an enlarged integral over z_i and λ_j but a single singularity surface given by the denominator in Eq. (12). Therefore we can apply Eq. (8). This gives Eq. (10) with the λ's all non-zero. A zero λ corresponds to replacing a pinch by an end-point in the corresponding λ-integration.

If the singularity is due to trapping the hypercontour in the edge given by $B_k = 0$, $k = 1,\dots,p$, the condition is found to be

$$\lambda_j \Sigma_j = 0 \qquad j = 1,\dots,m \; ;$$

$$B_k = 0 \qquad k = 1,\dots,p \; ;$$

$$\sum_j \lambda_j \frac{\partial \Sigma_j}{\partial z_i} + \sum_k \mu_k \frac{\partial B_k}{\partial z_i} = 0, \qquad i = 1,\dots n \; . \tag{13}$$

The μ's are essentially Lagrange multipliers taking into account the restraints imposed by the edge.

3. The Landau Equations

These equations were first derived as giving the location of the singularities of Feynman integrals. We shall see that they have a wider significance in the context of S-matrix theory.

Consider first the Feynman integral associated with a diagram with n lines:

$$\int_{-\infty}^{\infty} d^4k_j \int_0^1 d\alpha_i \frac{\delta(\Sigma\alpha_i - 1)}{[\Sigma\alpha_i(q_i^2 - m_i^2)]^n} \; , \tag{14}$$

where k_j are the loop momenta, and α_i, q_i, m_i are the Feynman parameter, momentum and mass, associated with the i-th line.

The k_j integrations stretch from $-\infty$ to $+\infty$ and are effectively topologically closed. Therefore they will only have pinches. There can be either end points or pinches in the α-integrations.

Writing

$$\psi \equiv \Sigma\alpha_i(q_i^2 - m_i^2) \quad , \qquad (15)$$

the conditions for a singularity are

$$\psi = 0 \quad ;$$

$$\frac{\partial\psi}{\partial k_j} = 0, \qquad j = 1,2,\ldots;$$

$$\alpha_i \frac{\partial\psi}{\partial\alpha_i} = \alpha_i(q_i^2 - m_i^2) = 0 \quad , \qquad i = 1,\ldots,n \quad . \qquad (16)$$

These are the Landau equations.

The Landau equations arise in S-matrix theory as the location of the singularities of unitarity integrals. The bubbles in the diagrammatic representation of these integrals correspond to scattering amplitudes $A^{\pm}$ which have singularities. These singularities, together with the phase space boundaries of the integrals generate further singularities of the unitarity integrals. Because of close connection (to be investigated later) between the singularities of the unitarity integrals and the singularities of the amplitudes themselves, these new singularities generated in this way can be fed back into the bubbles to generate further singularities; and so on.

As a simple example of this mechanism, we first consider the two-particle unitary integral represented schematically in Fig. 2. f_1 and f_2 are scattering amplitudes on appropriate sheets. They will have poles in the crossed channel. We denote the corresponding momenta by q_1' and q_2'. The integration is over the loop momentum k subject to the conditions

$$B_1 = q_1^2 - m^2 = 0 \quad ;$$

$$B_2 = q_2^2 - m^2 = 0 \quad . \tag{17}$$

Fig. 2

The problem of finding the singularity associated with this integral is formally the same as that of finding a pinch in an edge by $B_1 = B_2 = 0$. Therefore, by Eq. (11), its equation is given by

$$q_1^2 - m^2 = q_2^2 - m^2 = 0 \quad ;$$

$$\lambda_1(q'^2_1 - m^2) = \lambda_2(q'^2_2 - m^2) = 0 \quad ;$$

$$\lambda_1 q'_1 + \lambda_2 q'_2 + \mu_1 q_1 + \mu_2 q_2 = 0 \quad . \tag{18}$$

With λ_1, λ_2 non-zero this is just the Landau equation for the leading singularities of Fig. 3, which in this context we call a Landau diagram rather than a Feynman diagram. Notice that crossing symmetry, producing the poles in q'^2_1 and q'^2_2 is an essential ingredient in this argument. It is this requirement which gives a relativistic theory its richness and complication.

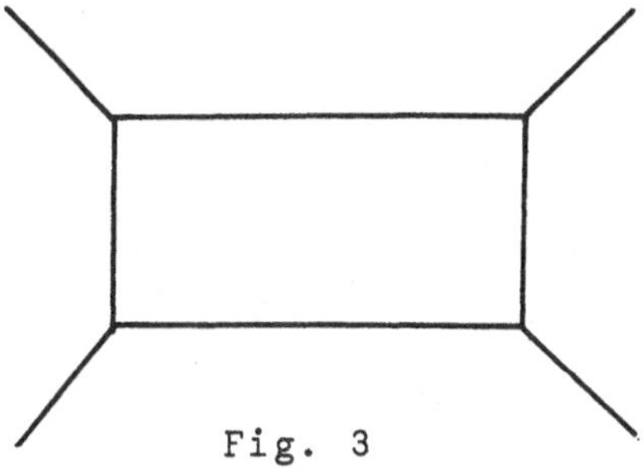

Fig. 3

In fact it can be shown that this procedure is completely general and is perfectly mirrored by the diagrammatic notation of Landau-Cutkosky diagrams: e.g.

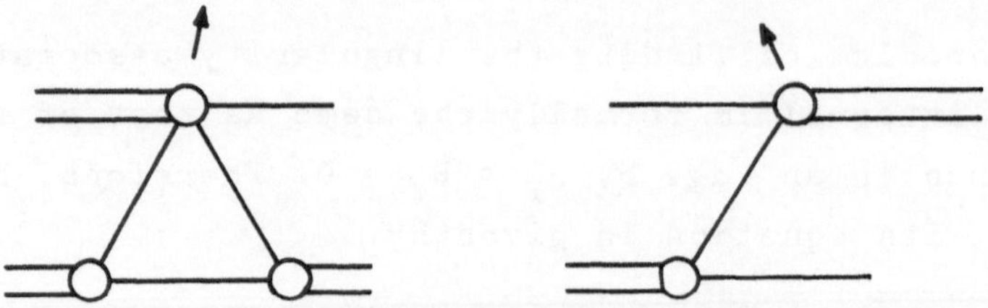

leads to a
singularity of the integral
corresponding to

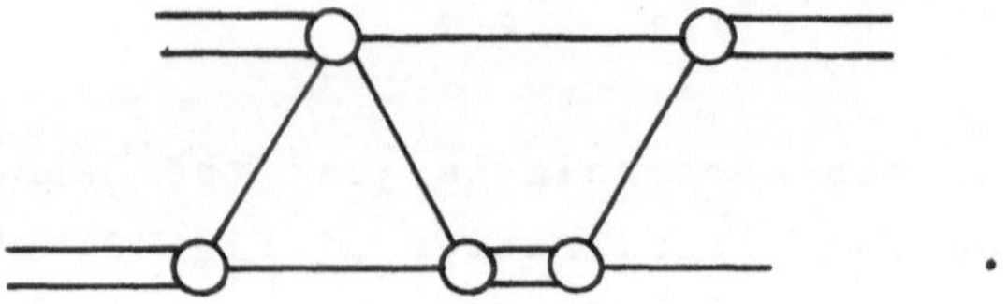

It might seem that in this way singularities of an integral could only correspond to diagrams with extra lines but in fact it is possible to contract out closed loops of lines from the phase space integral, so that (19) could also lead to the singularity corresponding to

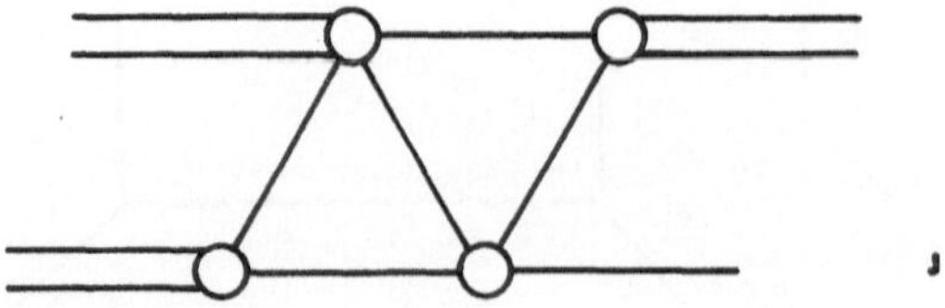

(20)

where the internal double lined loop has been contracted out.

However these methods only so far answer the question of where the integrals may be singular. Whether they actually are will depend on whether the contours are actually trapped or not.

4. Normal Thresholds and Single Particle Poles

The first S-matrix singularity to be discussed was the normal threshold, which occurs when a new channel becomes possible in a scattering process. It provides a simple example of some of the methods we shall use in later more complicated cases. We discuss the threshold at $9\mu^2$ in the $2 \to 2$ particle amplitude which corresponds to the opening up of the 3-particle channel. Because of this new possibility unitarity changes its form at $9\mu^2$:

$4\mu^2 \leq s < 9\mu^2$ (21a)

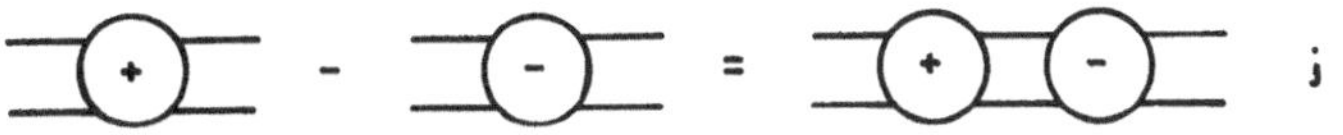

$9\mu^2 \leq s < 16\mu^2$ (21b)

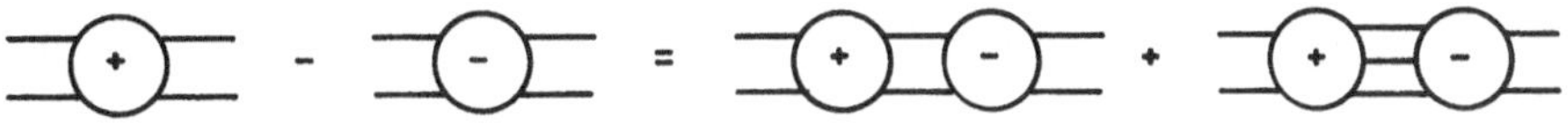

If we are to compare these two relations we must do so at the same value of s, that is we must analytically continue (21a) into the region $s > 9\mu^2$. Our path of continuation must avoid the singularity at $s = 9\mu^2$ either by going above or below this point in the complex s-plane. This may or may not take the ⊕ or ⊖ amplitudes into themselves. If continuation above is the correct way to link the ⊕ amplitudes in the two regions then continuation below is the correct path to link the two ⊖ amplitudes, since their paths of continuation must be related by complex conjugation. However we must choose the same path of continuation for all terms in (21a) if the equation is to remain valid. In fact we choose to continue it below $9\mu^2$ so that on continuation ⊕ goes onto an unphysical sheet below the $9\mu^2$ cut (see Fig. 4), where we denote it by ⓘ,

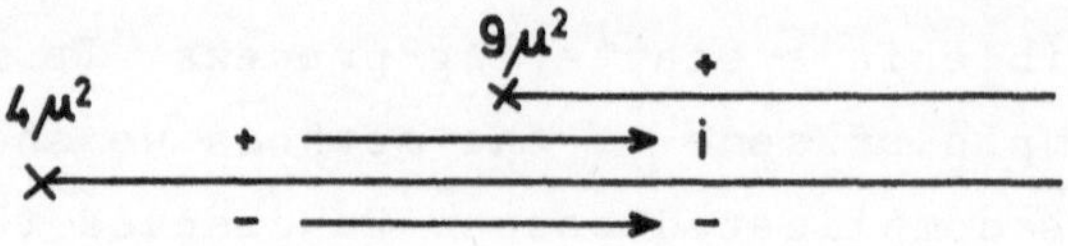

Fig. 4

so that (21a) becomes

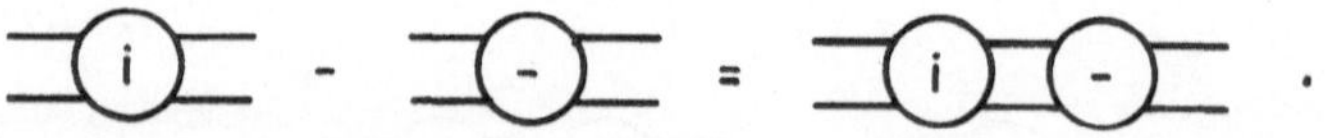

(21c)

Subtracting (21c) from (21b) yields, after some tidying

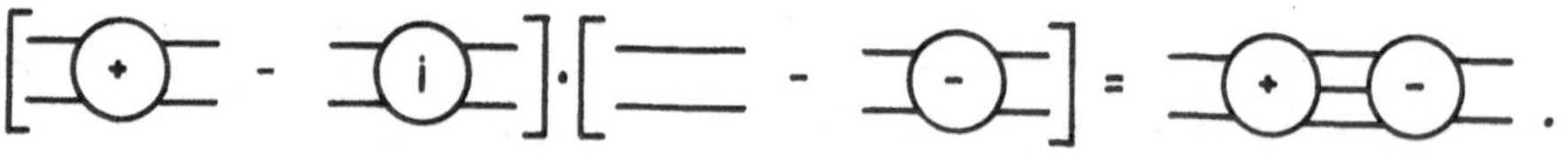

(22)

According to (21c) the inverse of the second factor on the left hand side of (22) is

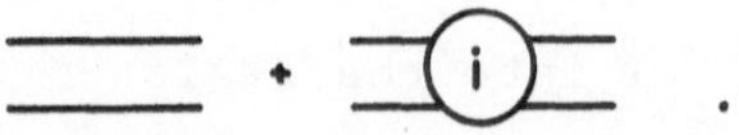

(23)

Use of this inverse yields

(24)

where

(25)

We interpret (24) as determining the discontinuity of ⊕ round its $9\mu^2$ threshold. Notice this is not the extra term appearing in (21b) but differs from it by replacing ⊖ by ⓘ.

The important general notion that this example illustrates is that we can isolate the discontinuity associated with a physical region singularity by comparing unitarity above the singularity with a suitable analytic continuation of unitarity below it.

To obtain more interesting singularities than normal thresholds we must have the courage to tackle multi-particle amplitudes. One of the first encountered is the single particle pole

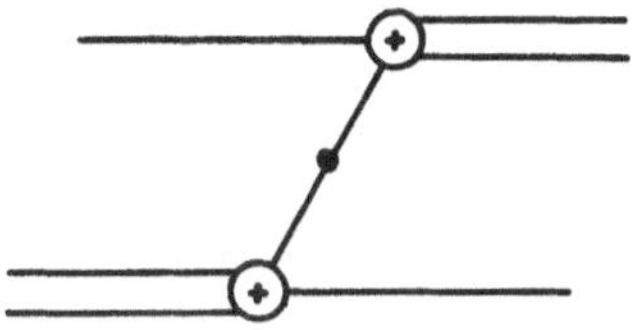
, (26)

whose presence is suggested by the unintegrated δ-function on the right hand side of the unitarity relation (3) in the term

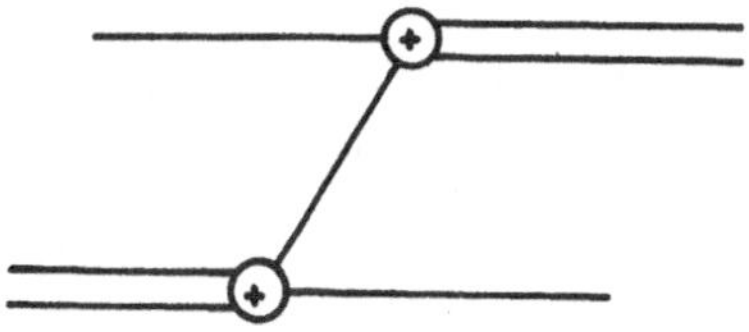
(27)

The term (27) is a strange one for if $q^2 \neq \mu^2$ (where q is the momentum in the single internal line) then there is no contribution from (27). Accordingly we shall compare unitarity at $q^2 = \mu^2$(with (27)) with the continuation of unitarity to $q^2 = \mu^2$ from a nearby point (which will lack (27)). We must decide from which direction to approach $q^2 = \mu^2$, choosing in fact to do so from below, $q^2 \to \mu^2 - i\varepsilon$, which we assume is correct for $\ominus$ but takes $\oplus$ onto the wrong side of the singularity, where we denote it by ⓘ. We notice also that the presence of the singularity in the larger bubbles of

(28)

will not be affected by the phase space integrations so that, while on continuation the first term of (28) is unchanged, the second will become

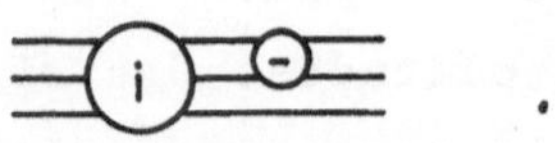
(29)

These are the only terms in which the singularity under discussion would be unaltered by integration.

Subtracting then the continued unitarity relation from the unitarity relation at $q^2 = \mu^2$ one obtains

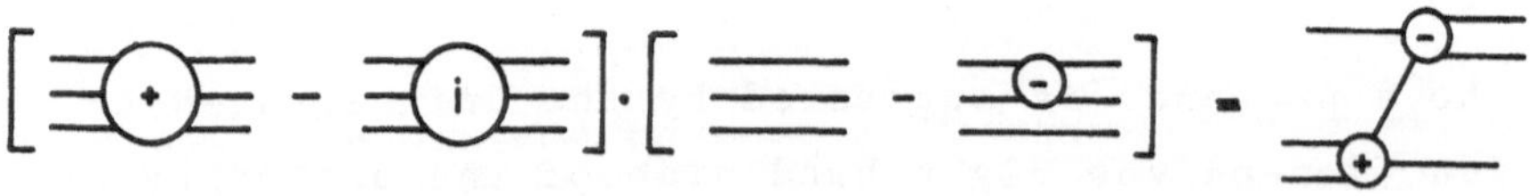
(30)

Two particle unitarity provides an inverse for the second factor on the left hand side of (30):

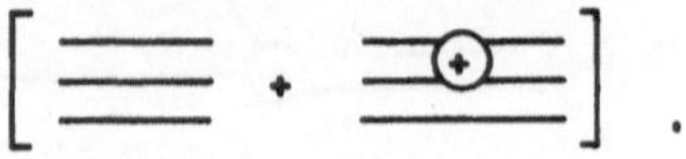
(31)

Multiplication by (31) and the use of two-particle unitarity again on the right hand side of the equation finally yields

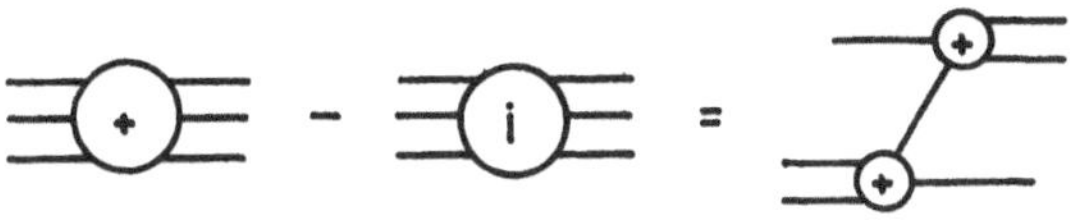

; (32)

i.e. the discontinuity is a δ-function so that the singularity is indeed the pole (26) with (+) bubbles at each end.

5. Physical Region Unitarity Integrals

The singularities discussed in the last section were all related to the appearance of extra terms in the unitarity equations. However, most physical region Landau singularities do not manifest themselves in this way. Before we can discuss these more subtle cases we must consider some simple ideas about unitarity integrals with real integration momenta.

The first mathematical tool we need is a criterion for the singularity of such a unitarity integral. I shall illustrate its nature by considering a single dimensional integral:

$$I(x) = \int \frac{dk}{(S_1 + i\varepsilon_1)(S_2 + i\varepsilon_2)} , \quad (33)$$

where the S_i are real functions of x and k. The $i\varepsilon_i$ are intended to indicate the limits in which the corresponding poles approach the real k-axis: they correspond to the prescriptions associated with ⊕ and ⊖ bubbles in S-matrix theory. Equation (33) will give a singularity when the two poles coincide. Its equations are

$$S_1 = 0 \quad ; \quad S_2 = 0 \quad ;$$

$$\alpha_1 \frac{\partial S_1}{\partial k} + \alpha_2 \frac{\partial S_2}{\partial k} = 0 \quad ; \tag{34}$$

where in our one-dimensional case the third equation is not a condition but serves as a definition (to within an overall factor) of α_1 and α_2. Equation (34) will only give a singularity if the coinciding poles actually trap the (real) k-contour , i.e. if they approach the real axis from opposite sides. The direction of approach is dertermined by ε_i. When this is non-zero the pole is displaced by an imaginary part $i\kappa_i$ given by

$$\frac{\partial S_i}{\partial k} \kappa_i + \varepsilon_i = 0 \ . \tag{35}$$

From (34) and (35) we see that the condition that the poles approach the real axis from opposite sides is that $\varepsilon_i \alpha_i$ has a sign independent of i. This condition generalizes to any number of real singularity surfaces with associated prescriptions and any number of integration variables. It is simple in form because we are dealing with a real contour of integration which is undistorted before the pinch forms.

When we continue the integral past a singularity we may still have an undistorted contour, or we may not, according to the way in which we have made the continuation. This is illustrated below:

Fig. 5

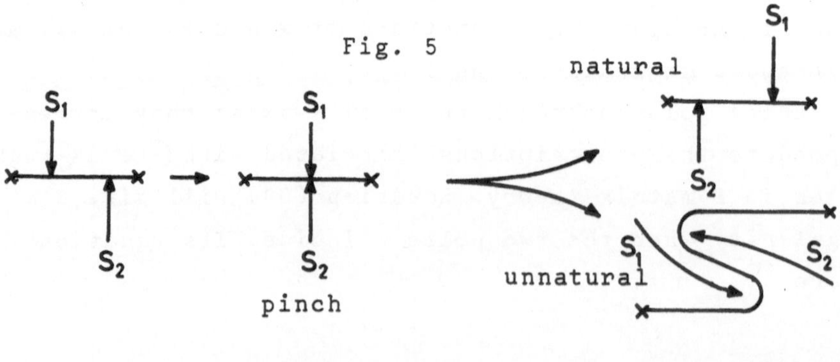

We call the continuation which leaves an undistorted contour the natural continuation. It can be determined in the following way: Consider

$$L(x) \equiv \alpha_1 S_1 + \alpha_2 S_2 \quad , \tag{36}$$

with the k,α dependence in (36) removed by satisfying the first and third of equations (34) and prescribing the value of α_2. Then L = 0 gives the singularity. Consider now how the singularity moves when S_i is changed by an amount $i\varepsilon_i$ (by changing, say, some parameter in S_i such as a mass). The implicit changes in α_i and k are irrelevant because

$$\frac{\partial L}{\partial \alpha_1} = 0 = \frac{\partial L}{\partial k} \quad ,$$

(from (34).) Thus the change in x is given by

$$\frac{\partial L}{\partial x}\, dx + i\Sigma\alpha_i\varepsilon_i = 0 \quad . \tag{37}$$

Let us define a variable η by

$$d\eta = \frac{\partial L}{\partial x}\, dx \quad , \tag{38}$$

so that locally $L \gtreqless 0$ corresponds to $\eta \gtreqless 0$, and the movement of the singularity is given by $d\eta = -i\Sigma\alpha_i\varepsilon_i$. Changing S_i by $i\varepsilon_i$ and keeping η real would give an undistorted contour. Thus if instead we make η imaginary in the opposite direction to the movement of the singularity we shall also keep the contour undistorted, see fig. 6.

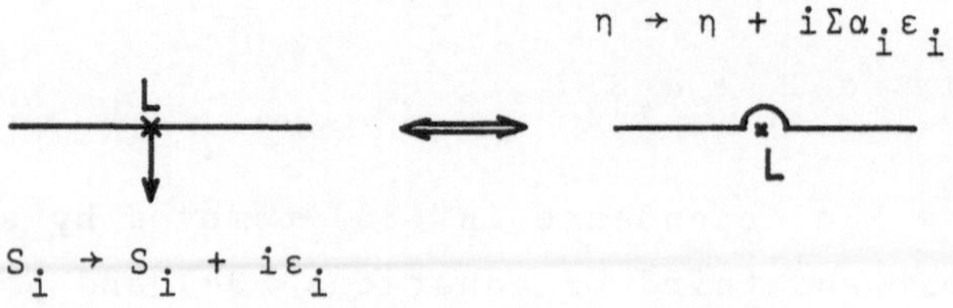

Fig. 6

Thus the natural distortion is given by going from $\eta < 0$ to $\eta > 0$ via $\eta + i\varepsilon_{nat}$ where ε_{nat} has the common sign of $\alpha_i\varepsilon_i$. Again this generalizes. In the case of several external variables x, η becomes a single variable measured normal to the Landau curve.

Finally we must be able to calculate the discontinuity on encircling the singularity. We can get this by considering the difference between the natural and unnatural continuations. Let us now suppose that S_1 and S_2 are singularities trailing cuts, rather than poles. Then the previous pictures lead to an expression for the discontinuity which is an integral joining S_1 to S_2 whose integrand is the discontinuity of the original integrand about the S_1 and S_2 singularities. Notice that as we tend to the singularity at $L = 0$, $S_1 \to S_2$ and the region of integration in the discontinuity expression vanishes. This is an example of what our topological friends call a vanishing cycle.

Again this result generalizes, though the generalization is far from trivial. The general picture which emerges is that the discontinuities of integrals are formed by compounding discontinuities of the integrand and integrating over vanishing cycles. Of course, keeping track of the signs is important. It turns out that this is made easier by using the special discontinuity we call "dif" which is the natural continuation minus the unnatural continuation, rather than using a clockwise or counter clockwise convention.

Finally, as far as integrals are concerned, we notice that in unitarity integrals we have not only singularities S_i by also constraints D_i (corresponding, for example, to mass shell conditions in phase space integrals). These D_i play a similar role to the S_i in generating Landau singularities but we do not have a condition on the corresponding parameters δ_i equivalent to the $\alpha_i\varepsilon_i$ condition. This is because they are always operative constraints rather than trapping singularities.

6. The Triangle Singularity

We are now in a position to discuss the work of Landshoff and Olive on the singularity

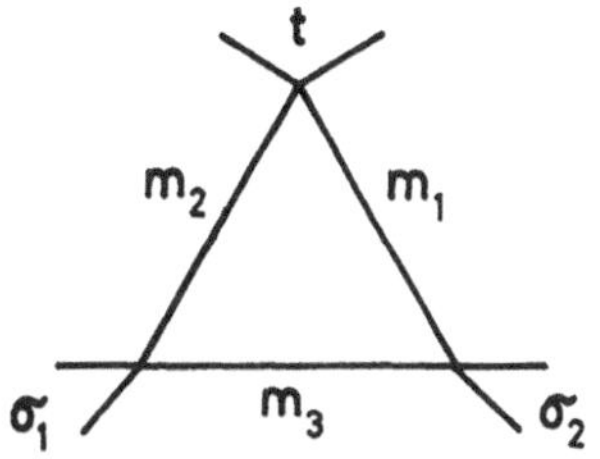

of

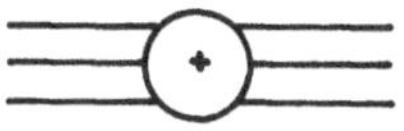

,

taking t fixed and negative (momentum transfer). Part of the Landau curve L lies in the physical region and looks like Fig. 7, where we have labelled the signs of the Feynman parameters on the different arcs of the curve.*

---

* To avoid degeneracies we are taking $m_1 \neq m_2$ but will neglect to represent this explicitly when we write unitarity equations. No essential difficulty is caused by this pedagogically convenient arrangement.

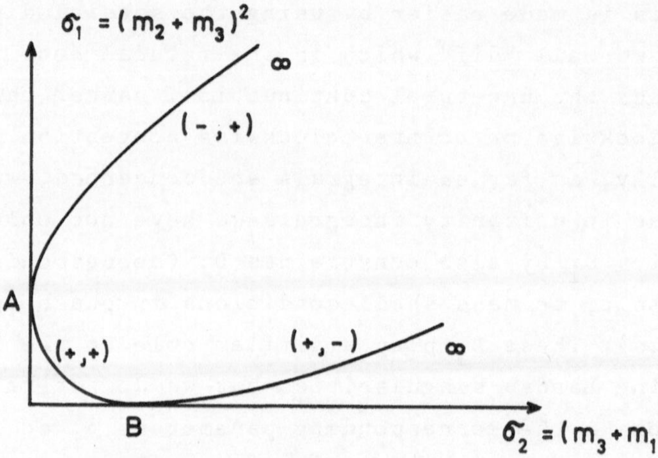

Fig. 7

Physical unitarity below the four-particle threshold is

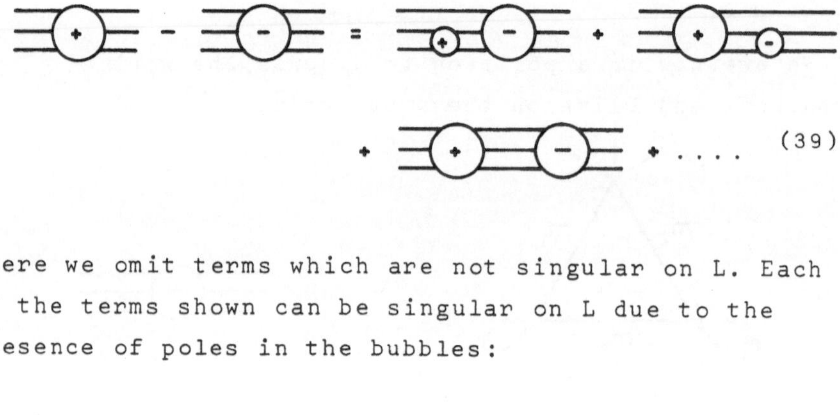

(39)

where we omit terms which are not singular on L. Each of the terms shown can be singular on L due to the presence of poles in the bubbles:

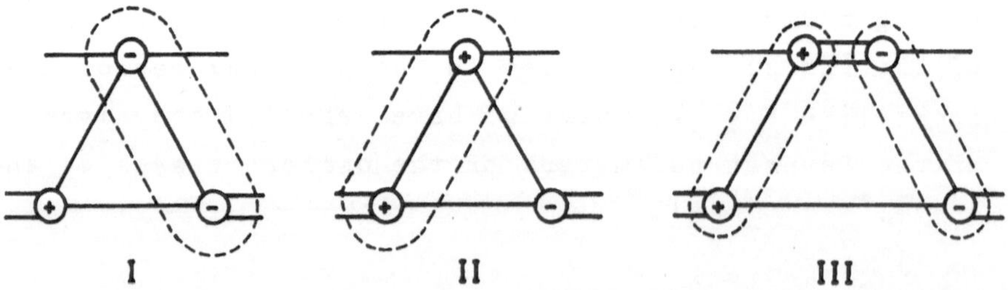

(40)

In III the upper loop must be contracted out. The pole terms have $\pm i\varepsilon$ associated with them according as they occur in $\textcircled{\pm}$ bubbles. Then according to the $\alpha_i\varepsilon_i$ criterion, I and II are singular on the whole of L, whilst III is singular only on ∞A, and B∞ . This refers to the unitarity integrals; according to perturbation theory we expect that the amplitude itself is singular only on the positive α arc, AB. We now see how this comes about.

The unitarity relation (39) holds both inside and outside L and so shows no manifest difference in the two regions. However, because of the singularities on L in I, II and III, the unitarity relation analytically continued from outside L to inside L does not coincide with the unitarity relation inside L. From the fact that both relations must be true we can extract non-trivial consequences.

We suppose the correct continuation for $\textcircled{+}$ across L is given by a $+i\varepsilon$ prescription in η, the variable normal to L which is negative outside and positive inside L. Then $-i\varepsilon$ in η will be the correct prescription for $\textcircled{-}$.

When we continue (39) across L the contours of integration in I, II and III will be distorted if the continuation is in the unnatural sense. The difference between the continued integral and the apparently similar integral in (39) inside L, which has an undistorted contour, will be just the corresponding discontinuity.

On the other hand,the two integrals will be identical if the continuation is made in the natural sense. According to the $\alpha_i\varepsilon_i$ rule the natural sense in η is given by

	∞A	AB	$B\infty$
I	$+i\varepsilon$	$-i\varepsilon$	$-i\varepsilon$
II	$+i\varepsilon$	$+i\varepsilon$	$-i\varepsilon$
III	$+i\varepsilon$	non-singular	$-i\varepsilon$

We call this effect the generation mechanism.

One further effect is present on continuation. The terms I and II may also be singular on L because of the presence of the complete singularity in the large bubble of each expression. This bubble is continued across L in the same sense as that chosen for (39) as a whole. We call this the regeneration mechanism*.

We now continue (39) across L along a route with $\eta - i\varepsilon$, ⊖ will continue into itself, but ⊕ will be continued onto the wrong side of L on a sheet where we denote it by ⓘ. Thus (39) becomes

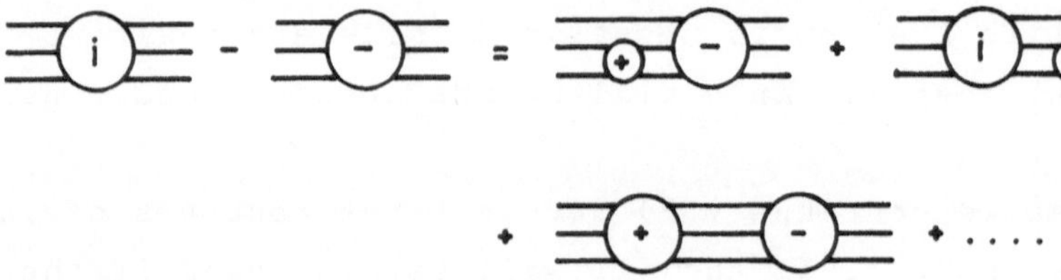

(41)

where in addition the contours of integration in the first three terms on the right hand side will be distorted if the $-i\varepsilon$ continuation is unnatural (i.e. in I, II, III across ∞A; in II only on AB; in none on $B\infty$). Subtracting (41) from (39) inside L, rearranging some terms, and correctly evaluating the discontinuities when contours are distorted in (41), yields

---

+ If explicit extra terms in unitarity were present we should call them the explicit mechanism.

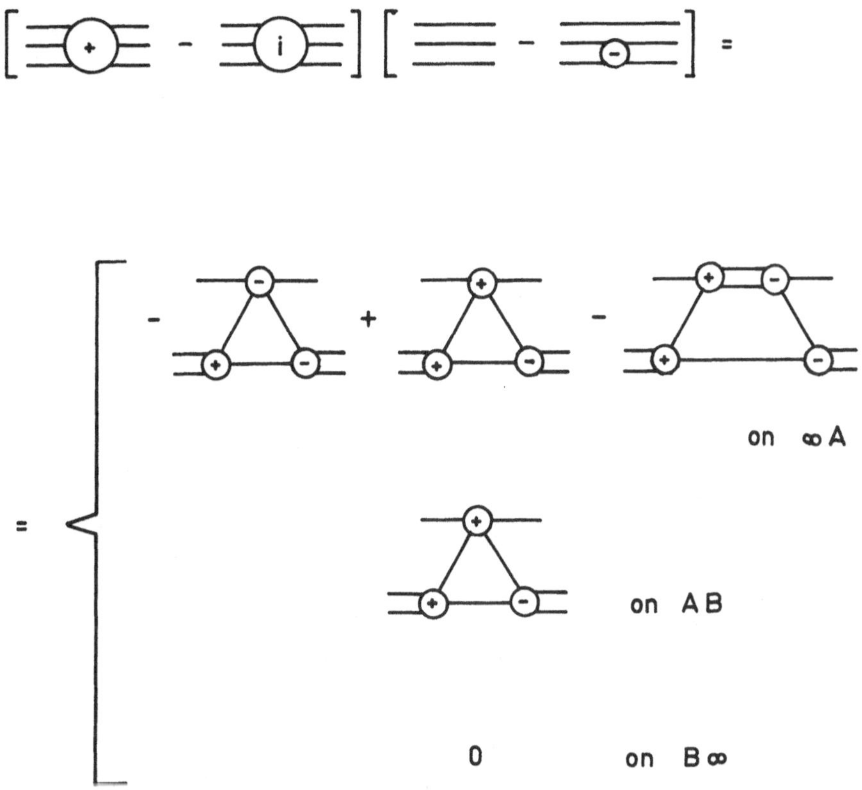

(42)

Two-particle unitarity for 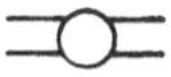 makes the total contribution to the right hand side of (42) zero on ∞A. It also provides an inverse for the second factor on the left of (42):

(43)

use of which then yields

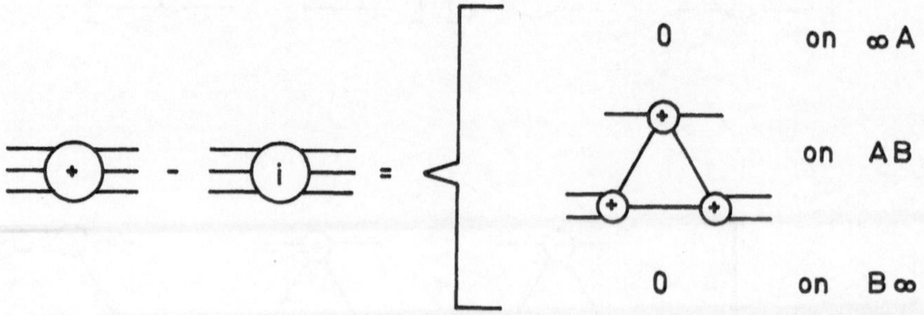

(44)

Thus we see that $\oplus$ is non-singular except on AB where its discontinuity is given by the Cutkosky formula.

7. Conclusion

The methods illustrated in the preceding section prove capable of generalization to show that all physical region singularities occur only on the positive α arcs of Landau curves and their discontinuities are given by Cutkosky integrals (see ref. 7). The problem of singularities outside the physical region still remains.

References

An account of the properties of Feynman integrals and the foundations of S-matrix theory can be found in "The Analytic S-Matrix" by R. J. Eden, P. V. Landshoff, D. I. Olive and J. C. Polkinghorne (Cambridge University Press 1966), which gives full references to the original literature. Further developments on the singularity structure of the S-matrix are to be found in the following papers:

1. P. V. Landshoff and D. I. Olive, J. Math. Phys. 7, 1464 (1966).
2. P. V. Landshoff and D. I. Olive, J. C. Polkinghorne, J. Math. Phys. 7, 1593 (1966).
3. P. V. Landshoff, D. I. Olive, J. C. Polkinghorne, J. Math. Phys. 7, 1600 (1966).
4. M. J. W. Bloxham, Nuovo Cim. 44A, 794 (1966).
5. P. V. Landshoff, J. Math. Phys., to be published.
6. J. K. Storrow, Nuovo Cim. to be published.
7. M. J. W. Bloxham, D. Olive, J. C. Polkinghorne, in preparation.
   This last is a sequence of four papers determining the complete S-matrix physical region singularity structure.

AN ATTEMPT TO CALCULATE RADIATIVE CORRECTIONS TO A PURE FERMI DECAY[†]

By

Gunnar KÄLLEN

Department of Theoretical Physics
University of Lund, Lund, Sweden

Summary

A general survey is given of an attempt to estimate the effective value of the cut off in the radiative corrections to nuclear β-decay.

I. Introduction

At the Schladming meeting last year, I gave a preliminary account of an attempt to calculate the radiative corrections to nuclear β-decay with special emphasize on the influence of nucleon structure, in particular the nucleon form factors. During the last year the calculation has been considerably refined and several small effects which were neglected before have been taken into account. Without going into too many calculational details I will try this year, first, to give again a description of the general idea behind the calculation and,

† Lecture given at the VI. Internationalen Universitätswochen für Kernphysik, Schladming, 26 February-11 March 1967.

second, to discuss the significance of the approximations involved.

Historically, the first calculations of radiative corrections to ordinary β-decay were performed on the basis of a point particle model, i.e., all particles including the neutron and proton with strong interactions were treated as simple Dirac particles with no structure and no anomalous magnetic moment [1]. Such a straight forward approach has the draw back that it leads to a result which is logarithmically divergent and, therefore, has to be handled with the aid of an arbitrary cut off. As the problem involved appears to be a problem of pure electrodynamics, this is a rather surprising situation. We are used to the fact that low order perturbation theory calculations in electrodynamics always lead to well-defined and finite results for observable effects like the Lamb shift, the anomalous magnetic moment of the electron, hyperfine structure corrections etc. Further, the effects calculated this way agree very well with experiments. In a perhaps somewhat oversimplified way of talking, this is the whole basis for modern field theory and all its byproducts like dispersion theory, S-matrix theory etc. Of course, various attitudes are possible in this situation. On the one hand, one might say that if any basic theory like quantum electrodynamics fails in a case with observable consequences like the radiative corrections to ordinary β-decay, it is an indication of some basic inconsistency or failure in the formalism. Logically, such an attitude is quite permissible. However, if one wants to adopt this kind of philosophy, one will have to declare that the radiative corrections to β-decay are a very significant phenomenon and that we here see a very basic failure of electrodynamics. In view of the brilliant successes of this theory in all other low energy phenomena and our great difficulty in finding any deviation from the theoretical predictions of electrodynamics up to very high energies, one can't help having the feeling that it might be better to look for an alternative explanation of the logarithmic divergence in β-decay and to try to blame this effect not on any basic principle but rather on some kind of physically unjustified approximation in the detailed calculation.

One thing which comes to mind here is the fact that the two nucleons involved in the β-decay are strongly interacting particles. Therefore, the virtual photon which enters in the calculation may very well interact not with the point particle nucleon but, rather, with a virtual π- or K-meson which in its turn is strongly coupled to the nucleons. The net effect of this coupling between the nucleons and mesons can be described by saying that the nucleons acquire a structure. This structure is presumably only important for reasonably high virtual momentum transfers. Therefore, it can be neglected in most processes where a calculation based on point particles gives a finite result. However, when this is not the case as in our problem, it means that large virtual momentum transfers are significant for the effect under consideration and, therefore, that the structure of the nucleons cannot be neglected. Consequently, the main modification of the standard calculation which is necessary becomes the introduction of the strong interaction effects via nucleon structure.

II. The Two Basic Problems

As no reliable theory exists which allows us to calculate the effect of the strong interactions, we must rely on the experimental information available. The best experiments which give us any insight into quantities like an effective radius of the proton etc. are the data on electron-nucleon scattering. As is very well-known, these scattering data can be summarized in terms of two form factors for the proton and two form factors for the neutron. The experimental data available refer to the case when the incoming and the outgoing nucleons are on the mass shell (at least approximately as the neutron may be bound in a deuteron) and that the photon is space-like. In a similar way one could imagine measurements, even if they do not exist today, where form factors with the photon and one of the nucleons on their mass shells but with the other one off the mass shell could be measured. However, a form factor where two of the particles are off the mass shell does not appear to be directly accessible to experimental investigations. On the other hand, if a form factor is introduced directly in a standard Feynman integral, we have to integrate over all momentum space and it appears that a rather extensive knowledge of the form factors with only one particle on the mass shell would be necessary. Therefore, it appears rather hopeless to try to estimate, in any reliable way, the influence of nucleon structure on the radiative corrections to β-decay because of our lack of knowledge of form factors off the mass shell. Also, it is not evident that all strong interaction effects can be described in terms of form factors alone.

Another question of principle which appears here is concerned with the fact that the electron has no strong interactions. An analysis of the divergence which appears in the standard formula for the radiative corrections to β-decay shows that the coefficient in front of the logarithm of the cut off energy is a sum of three terms, one coming from a "vertex part" where both the electron and the strongly interacting particles contribute, one coming from a "self energy diagram" of the proton, and a third part coming from a self energy diagram of the electron. (To avoid misunderstanding, it should perhaps be mentioned that the actual infinity we are concerned with here is not the infinite self mass which is handled by standard renormalization techniques but, rather, the field operator renormalization which is obtained from the self energy diagram.) The first two of these terms will certainly be modified by the strong interactions and, hopefully, made finite, while the contribution from the electron field operator renormalization is not influenced by strong interactions. In view of this it appears to be a rather hopeless task to try to obtain any finite value for the radiative correction to β-decay because of possible modifications due to strong interactions.

The first point which we should like to make in this connection is that the field operator renormalization infinity mentioned in the previous paragraph is not a gauge invariant quantity. Therefore, it cannot have any absolute significance. Consequently, it is tempting to try to formulate an approximation scheme in a gauge which deviates from the standard Gupta-Bleuler gauge in quantum electrodynamics in such a way that the field operator renormalization factor for a point particle is finite. Actually, a large number of such gauges exists already in the literature [2]. When we then introduce form factors in the vertex part and in the proton self energy diagram, it is at least possible that the total result of the calculation will be finite. Obviously, a calculation of this kind can never be exactly gauge invariant. It gives manifestly different results in gauges where the renormalization constant of the electron field operator is finite and where this constant is infinite. However, we can at least require that the result of the calculation must be the same in all gauges where this renormalization constant is finite. This is what we should like to call a restricted gauge invariance of our formalism. If this can be achieved it will be one possible way of avoiding the infinity in the electron self energy diagram which otherwise appears to be a stumbling block in this connection.

Next comes the question about contributions from off shell form factors. We propose to handle this problem in the following way. As is well-known from the historical development of modern field theory, the usual Feynman diagram technique is not the only way one has of writing perturbation theory. Actually, a historically older approach [3] gave expressions where all terms which are nowadays combined into one Feynman integral were written as a sum of several contributions involving δ-functions and principle value denominators. Evidently, as both approaches are basically straight forward perturbation theory they must be mathematically equivalent. Indeed, if one trys to calculate the contribution from a Feynman integral using residue calculus it is easy to arrive at the older form of perturbation theory. Each pole in the rational denominator of the Feynman diagram gives a δ-function contribution corresponding to one of the internal particles being on its mass shell. From our point of view this last feature is particularly interesting. Let us assume that the older version of the formalism is in a sense more basic than the Feynman version. It then becomes possible for us to introduce form factors in such a way that one of the intermediate particles is always on its mass shell. Such form factors are, as has been pointed out above, directly observable in contradistinction to the general off shell form factors. Therefore, if we write the formalism in this way we can calculate the radiative corrections to β-decay including the influence of the strong interactions in terms of form factors which either have been measured or, in principle, could be measured. In another language, this kind of approach is intimately related to dispersion theory methods.

III. General Method of Calculation

The ideas indicated above can be carried through in some mathematical details. We have no intention of entering into the calculations here. Some of the arguments were given last year at the Schladming meeting, even if in a somewhat primitive shape. The complete calculation has been submitted to the Nuclear Physics and is expected to appear soon [4]. However, a few points could perhaps be mentioned. We write the weak interaction Lagrangian as a product of two factors, one being the weak leptonic current and the other the hadronic current.

$$\mathcal{L}_W = - \ell_\lambda(x)[j_\lambda^\beta(x) + j_\lambda^5(x)] + \text{h.c.} \quad . \tag{1}$$

The leptonic current $\ell_\lambda(x)$ is given by the standard expression

$$\ell_\lambda(x) = \frac{g}{\sqrt{2}} \bar{\psi}_e(x)\gamma_\lambda \; \psi_\nu(x) \quad , \tag{2}$$

while the hadronic vector and axial vector currents also involve strong interactions and, therefore, cannot be written down explicitly. The transition probability per unit time for the β-decay of the neutron to a proton, an electron and an antineutrino can now be written with the aid of the elementary formula

$$\frac{\delta w}{\delta \tau} = 2\pi \sum_{\substack{\text{final} \\ \text{states}}} \delta(E_i - E_f) \left| \int \langle p,e,\bar{\nu}| \; \mathcal{L}_W(x) |n\rangle d^3x \right|^2 \quad . \tag{3}$$

Therefore, the basic problem is to calculate the indicated matrix element of the weak interaction Lagrangian. For this purpose the first factor involving $\ell_\lambda(x)$ is of great help, as this quantity is made up of the electron field and the neutrino field, none of which has any strong interactions. Indeed, the neutrino field has no interaction at all except for the β-interaction and, therefore, can be treated as a free field in Eq. (3). The electron field, however, has an electromagnetic interaction, but this is an effect which can be reliably handled with the aid of perturbation theory. Consequently, if we analyze the matrix element of the weak interaction Lagrangian with the aid of an expansion in terms of a complete set of intermediate states $|z\rangle$, we find

$$\langle p,e,\bar{\nu}|\,\mathcal{L}_W(x)\,|n\rangle =$$

$$= - \sum_{|z\rangle} \langle p,e,\bar{\nu}|\,\ell_\lambda(x)\,|z\rangle\langle z|\,j_\lambda^\beta(x) + j_\lambda^5(x)\,|n\rangle \;. \qquad (4)$$

IV. One-Nucleon Intermediate States

The first factor in Eq. (4) involving the leptonic current alone decides which states $|z\rangle$ actually contribute in the expansion. Starting with those states which have as low a mass as possible, one finds that states with one nucleon, one nucleon and one photon and states with one nucleon and one electron-positron pair contribute to lowest order in the fine structure constant α . As an example, let us consider the first of these states, viz. states $|z\rangle$ with one nucleon,and let us assume that this nucleon is not the same as the proton in the final state. Using standard expressions for a perturbation theory expansion of the electron operator [5] one finds that this matrix element of the leptonic current can be expressed in terms of explicitly known quantities like γ-matrices, the solution to the free particle Dirac equation etc. and a matrix element between the initial proton and the proton in the intermediate state of the electromagnetic current. The intuitive reason for the appearance of such a factor is that the electron is coupled to the protons only through the electromagnetic field. Therefore, any matrix element between states with different nucleons must involve the current of the electromagnetic interaction. However, this electromagnetic current is essentially given by the form factors measured in electron-proton scattering. Therefore, even if this quantity cannot be calculated theoretically, it can be evaluated numerically using experimental data.

There remains the second factor in Eq. (4), viz. the matrix element of the vector and axial vector weak currents between the proton in the intermediate state and the neutron in the initial state. Assuming the conserved vector current hypothesis, the vector part of this matrix element can be related to the corresponding matrix element of the electromagnetic, isotopic vector current and, therefore, again to the proton and neutron form factors. Consequently, also this quantity is experimentally known. If we had a process where only the vector interaction were effective, we could, therefore, calculate the corresponding contribution to Eq. (4) reliably using only an expansion in the electromagnetic interaction and experimentally known form factors for the strongly interacting particles. At the first moment, one might be inclined to think that a pure Fermi transition which is known to involve only the vector interaction when radiative corrections are neglected, would be an ideal object for our calculation. However, that is not the case. Also in a pure Fermi transition one gets interference terms between the vector and axial vector part of the weak interaction when radiative corrections are taken into account. This happens essentially because the ordinary selection rules in the β-decay are the selection rules of angular momentum. However, the photon which appears implicitly in our calculation carries a spin and, therefore, the angular momentum selection rules are influenced. The net effect of this is the appearance of a contribution from the axial vector part of the β-interaction in Eq. (4) also in a pure Fermi transition.

Consequently, we are forced to discuss the matrix element of the axial vector current between a proton and a nucleon state. The main point to be made here is that such a matrix element is, in principle, observable and actually has already been observed, even if the accuracy involved is rather small. In experiments where a neutrino is scattered by a nucleon giving rise to an inelastic process with a proton and a lepton in the final state, one measures a cross section depending on all the form factors of the weak interaction under discussion here [6]. Experiments of this kind have been performed and have indicated that the axial vector form factor of the weak interaction hadronic current behaves qualitatively in the same way as the vector form factors. Therefore, and in principle, the complete expression in Eq. (4) stemming from intermediate states with one nucleon, can be expressed in terms of observable quantities.

V. Other Low Mass Intermediate States

Unfortunately, the other intermediate states in Eq. (4) cannot be expressed in terms of experimentally known quantities. However, they can be analyzed in the same way as the one-particle states. The main difference is that the form factors which enter refer to inelastic processes and stem from matrix elements of the electromagnetic current between a state with one proton and another state with another proton and a photon or an electron-positron pair. In principle, the first kind of matrix elements can be observed if one measures the cross section for the bremsstrahlung of a proton in a heavy external field. In practice, such a heavy external field would probably be a heavy nucleus like a uranium nucleus [7]. Assuming that experiments on this process with an accuracy to make them interesting for our purpose would be available one day, one could use such data to calculate the contributions from these intermediate states in the same way as we have handled the one-proton intermediate states above. However, no such data are available today, at least to our knowledge. Nevertheless, it appears reasonable to assume that the form factors which enter in such a calculation have the same qualitative behaviour as the form factors known today. Therefore, and to get an idea about the numerical significance of the strong interactions in the radiative corrections to β-decay, we have introduced an effective form factor for these matrix elements in our calculation. We assume the same analytic structure for this form factor as for the ordinary form factors but with a slope which is considered an unknown parameter. To simplify the formulae as much as possible only one new form factor is introduced roughly corresponding to the Dirac form factor for the one nucleon states. In principle, more form factors corresponding to anomalous magnetic moment terms etc. should also be considered. However, we feel that such a refinement can wait until more detailed experimental information is available. In the final result, the slope of the form factor has been put equal to the slope of the known proton and neutron form factors.

The intermediate states involving one nucleon and an electron-positron pair are then, in principle, treated in the same way. The form factor that enters here is analogous to the form factor in the previous case but with the photon on the mass shell replaced by a time like photon corresponding to the electron-positron pair. However, one finds by kinematical considerations that the effective mass of the electron-positron pair is very small compared to any reasonable strongly interacting particle mass (e.g. the π-meson mass) and, therefore, the photon in these matrix elements can be approximated as being on its mass shell [8]. Consequently, the same form factor appears here as in the previous case.

VI. High Mass Intermediate States

The intermediate states discussed until now are those which have the lowest mass and also those which appear explicitly in a perturbation theory treatment based on point particles. However, the electron is coupled to all charged particles, i.e., also to charged π-mesons, charged K-mesons etc. Therefore, when we sum over a complete set of intermediate states in Eq. (4), we should also consider contributions from states involving, e.g., one proton and one π-meson, possibly together with one photon or one electron-positron pair. All such states have an effective mass which is larger than the sum of the nucleon mass and the π-meson mass and, therefore, well separated from the effective masses of the states considered so far. If we write these contributions down in a way similar to the treatment indicated for the other states, we would obtain expressions involving matrix elements of the electromagnetic current between states with one proton and one proton and a π-meson (in the simplest case). One might possibly expect a π-meson nucleon intermediate state corresponding to the $N^*(1238)$ resonance to be of particular importance. Actually, the matrix elements of interest here are possibly available from data on the photon-production of this resonance. Therefore, they could perhaps be incorporated in our analysis. However, this has not been done so far, mainly with the motivation that the inclusion of these states would probably not change our final result very much because of the high virtual mass involved. Further, our result contains so many unknown parameters from the unknown form factors anyhow that a too refined treatment does not appear reasonable. Therefore, these states have been neglected as well as higher states involving K-mesons or a larger number of π-mesons etc. Clearly, this is not a question of principle but one has to stop the calculation somewhere and this appears a reasonable stage to break it off.

VII. Numerical Results

On the basis of the ideas sketched above it is possible to arrive at a definite number for the radiative corrections to an arbitrary β-decay. As precision measurements are here mainly of interest for pure Fermi decay, we have specialized the formalism above to that case. In this way we find the following numerical result

$$\frac{1}{\tau} = \frac{1}{\tau_o}\left[1 + \frac{\alpha}{\pi}\left(\frac{3}{2}\log\frac{M}{m} + \Delta + h(\varepsilon)\right)\right] \quad , \qquad (5)$$

$$\Delta \cong 13.7 \pm 1.1 \quad , \qquad (5.a)$$

$$h(\varepsilon) \approx -\frac{3}{2}\log(2\varepsilon) - 2.1547 \quad . \qquad (5.b)$$

Here, τ is the lifetime including the radiative corrections while τ_o is the same quantity without radiative corrections but including standard Coulomb corrections as well as screening effects, the effect of finite nuclear size etc. The mass of the nucleon has been denoted by M while m is the electron mass. Further, ε is the maximum kinetic energy for the electron expressed in units of the electron mass. The function $h(\varepsilon)$ is a complicated algebraic expression but is approximated by Eq. (5.b) for decays where ε is large. Actually, it is accurate to about 10 % if ε is larger than 2. The numerical value of the constant Δ in Eq. (5.a) depends on the form factors as discussed above. The error in this constant takes into account the experimental error of the axial vector form factor for the one proton states and also contains an estimate of the uncertainty in the unknown form factors involving the extra photon or the extra electron-positron pair. However, the systematic error because of the neglect of intermediate states with masses larger than the sum of the nucleon and π-meson mass is not included.

Using the result indicated in Eqs. (5) we can evaluate the radiative corrections to various Fermi transitions. In this case one finds the result summarized in Table 1.

Decay	ε	$h(\varepsilon)$	$\Delta\tau/\tau$ %	$(\Delta\tau/\tau)_{KS}$ %
$^{10}C(\beta^+)^{10}B^{**}$	1.74	-4.47	-2.1	-1.9
$^{14}O(\beta^+)^{14}N^*$	3.55	-5.31	-2.0	-1.6
$^{26}Al(\beta^+)^{26}Mg$	6.28	-6.06	-1.8	-1.4
$^{34}Cl(\beta^+)^{34}S$	8.73	-6.53	-1.7	-1.3
$^{42}Sc(\beta^+)^{42}Ca$	10.59	-6.80	-1.6	-1.2
$^{46}V(\beta^+)^{46}Ti$	11.80	-6.96	-1.6	-1.2
$^{50}Mn(\beta^+)^{50}Cr$	12.93	-7.09	-1.5	-1.2
$^{54}Co(\beta^+)^{54}Fe$	14.14	-7.22	-1.5	-1.1

Table 1

Radiative corrections for some pure Fermi decays. $\Delta\tau = \tau - \tau_o$.

The last column here shows the value one obtains using the point particle formula and arbitrarily putting the cut off equal to the nucleon mass [9]. We note that the numerical results in the two columns differ by about 0.4 % for the lifetime which corresponds to about 0.2% in the coupling constant. This is a change which is essentially of the same order of magnitude as the experimental error today and, therefore, not very significant. At the same time it should be noted that the uncertainty of the constant Δ indicated in Eq. (5.a) gives an uncertainty in our numerical result for the lifetime of about 0.2 %, i.e. half the difference between the old and the new calculation. As a rough summary we can therefore say that there is no very significant difference between our result and the older procedure and we can consider our argument as a justification for the usual method of putting the cut off in the Kinoshita-Sirlin formula equal to the nucleon mass. Nevertheless, it must be emphasized that the actual numerical result we have presented here is somewhat uncertain because of the influence of form factors, the behaviour of which is either guessed or estimated on the basis of rather inprecise experimental information. Therefore, a future re-evaluation of our formula based on more accurate experimental data may give a different result. Consequently, the main conclusion is perhaps that our argument indicates that the appearance of the logarithmic infinity found in the standard radiative corrections to ordinary β-decay is not necessarily of basic significance but can very well be removed if strong interaction effects are taken properly into account.

References and Footnotes

1. Cf., e.g., T. Kinoshita and A. Sirlin, Phys. Rev. <u>113</u>, 1652 (1959). More complete references can be found in this paper as well as in ref. 4 below.
2. Cf., e.g., the "Landau gauge" used by Landau et al.

in a series of papers about the high energy behaviour of quantum electrodynamics. All these papers are reprinted in "Collected papers of L. D. Landau", Pergamon Press (1965).

3. Cf., e.g., J. Schwinger, Phys. Rev. 76, 790 (1949).
4. G. Källén, Nuclear Physics, to be published.
5. Cf., e.g., the proceedings of the Schladming meeting 1966.
6. J. Løvseth, Phys. Lett. 5, 199 (1963). Experimental results in M. M. Block et al., Phys. Lett. 12, 281 (1964) and J. K. Bienlein et al., Phys. Lett. 13, 80 (1964).
7. The bremsstrahlung in a proton-proton scattering process would be unsuitable for this purpose as the recoil of the target could not be neglected in such a reaction.
8. Cf. the calculations given in the 1966 Schladming proceedings, esp. on p. 211.
9. Cf. ref. 1.

QUANTUM FIELD THEORY AND GENERALIZED FUNCTIONS†

By

F. ROHRLICH

Department of Physics, Syracuse University
Syracuse, USA

In the following lectures I wish to review the present state of quantum field theory following the approach usually called "asymptotic quantum field theory". Since certain mathematical developments, especially from the theory of generalized functions, will be an important prerequisit, I shall proceed as follows. In order to avoid interrupting the development of the physical theory by mathematical asides, I shall first present several primarily mathematical topics which will then be used in the second part of these lectures for the development of the physical theory. The outline is therefore as follows:

A. Mathematical Preliminaries
   1. Generalized Functions
   2. Analytic Functionals and the Class Δ_Γ^n of Generalized Functions
   3. The Free Field Equation $K^N a = 0$
   4. Operator Derivatives
   5. The Operators P_{12} and Their Algebra

† Lecture given at the VI. Internationalen Universitätswochen f.Kernphysik,Schladming,26 February-11 March 1967.

A. Mathematical Preliminaries

1. Generalized Functions

In this lecture I want to define a class of analytic functionals which is a generalization of the well-known functions $\Delta(x)$, $\Delta_R(x)$, $\Delta_C(x)$, etc. of standard quantum field theory. The latter are actually not functions in the usual sense, but generalized functions. They can be given a precise mathematical meaning only concomitant with a space of sufficiently well-behaved functions, a test function space. <u>Generalized Functions</u> are continuous linear functionals T defined on a test function space Φ. Under suitable conditions these T also define a space. It is called Φ', the conjugate space to Φ.

Example 1: Let $\Phi = \mathcal{S}$, the space of all infinitely differentiable functions (in one dimension for the sake of this example) which fall off faster than any power. The space $\mathcal{S}'$ is the space of all tempered distributions. More precisely,

$$|x^n \frac{d^m}{dx^m} f(x)| \leq B_{mn} \qquad \forall f \in \mathcal{S} \; ; \; m,n > o \qquad (1.1)$$

Example 2: Let $\Phi = \mathcal{K}(a)$, the space of all infinitely differentiable functions which vanish outside $|x| \leq a$. The conjugate space of all continuous linear functionals on $\mathcal{K}(a)$ is denoted by $\mathcal{K}'(a)$. The union over a, $\cup\mathcal{K}(a) \equiv$

$\equiv \mathcal{D}$ is the test function space of Schwartz distributions. Its conjugate is $\mathcal{D}'$.

For $T \in \Phi'$ and $f \in \Phi$ one writes the functional as

$$T(f) = (T,f) \tag{1.2}$$

or in terms of a <u>formal</u> integral

$$T(f) = \int T^*(x)\, f(x)\, dx \tag{1.3}$$

The Dirac delta-function, for example, which belongs to $\mathcal{S}'$, is written

$$\delta(f) = \int \delta(x)\, f(x)\, dx \equiv f(o) \tag{1.4}$$

by definition, for all $f \in \mathcal{S}$.

If $T(x)$ in (1.3) is an ordinary function, $F(x)$ say, so that the integral has a meaning in the sense of Riemann or Lebesgue, $F(f)$ is of course defined. Thus, $T(f)$ with $T \in \Phi'$ can be regarded as a generalization of $F(f)$ for the case when $T(x)$ is no longer a function (as in the case of $\delta(x)$). Hence: "Generalized Functions".

Multiplication of T with (complex) numbers is clearly defined and one has for $\alpha,\beta \in C$, for $S,T \in \Phi'$, and $f,g \in \Phi$,

$$(\alpha T + \beta S, f) = \alpha^*(T,f) + \beta^*(S,f) \tag{1.5}$$

$$(T, \alpha f + \beta g) = \alpha(T,f) + \beta(T,g) \tag{1.6}$$

Thus (T,f) is linear homogeneous in f and linear antihomogeneous in T. The latter property is sometimes called antilinear. However, it is (1.6) which characterizes $T(f)$ as a <u>linear</u> functional.

The product $S(x)\,T(x)$ is not defined in general since one wants the associative law of multiplication to hold,

$$\int (S(x)\ T(x)) f(x)\ dx = \int S(x)\ (T(x)\ f(x))\ dx$$

and $T(x)\ f(x) \notin \Phi$ in general. Example: The step function

$$\Theta(x) = \begin{cases} 1 & x > o \\ 0 & x < o \end{cases} \qquad (1.7)$$

can be regarded as a distribution. Now $\delta(x)\ \Theta(x)$ is not defined because $\Theta(x)\ f(x) \notin \mathcal{S}$ for <u>all</u> $f(x) \in \mathcal{S}$. (It is $\in \mathcal{S}$ for <u>some</u> f).

The convolution product is defined by

$$(S*T,\ f) \equiv \int S(x)\ T(y)\ f(x+y)\ dxdy$$

and exists whenever one of the two generalized functions has bounded support.

Derivatives of $T \in \Phi'$ are defined to all orders by

$$((\frac{d}{dx})^n\ T,f) = (-1)^n\ (T,\ (\frac{d}{dx})^n\ f) \quad . \qquad (1.8)$$

This relation can be obtained from (1.3) by integration by parts.

The Fourier transform of T(x) can be defined as usual by

$$f(x) = \frac{1}{\sqrt{2\pi}} \int \tilde{f}(p)\ e^{ipx} dp \qquad (1.9)$$

$$T(x) = \frac{1}{\sqrt{2\pi}} \int \tilde{T}(p)\ e^{ipx} dp$$

$$(T(x),f(x)) = \frac{1}{2\pi} \int \tilde{T}^*(p)\ \tilde{f}(q)\ e^{iqx-ipx}\ dx\ dp\ dq =$$

$$= \int \tilde{T}^*(p)\ \tilde{f}(p)\ dp = (\tilde{T}(p),\ \tilde{f}(p)) \quad .(1.10)$$

It is clear, however, that the properties of f which

make it an element of Φ will in general not hold for $\tilde{f}$ so that $\tilde{f} \notin \Phi$. Only in the special case $f \in \mathcal{S}$ it is true that also $\tilde{f} \in \mathcal{S}$.

In that case therefore $T \in \mathcal{S}'$ implies $\tilde{T} \in \mathcal{S}'$ and the Fourier transform of a tempered distribution is again a tempered distribution.

Consider now $T \in \mathcal{K}'(a)$. Let the <u>support</u> of a function be defined as the complement of the set of points on which it vanishes. Then $f \in \mathcal{K}(a)$ has bounded support. Therefore $\tilde{f}$ is defined also for complex arguments,

$$\tilde{f}(p_1 + ip_2) = \frac{1}{\sqrt{2\pi}} \int_{-\infty}^{\infty} f(x)\, e^{-i(p_1+ip_2)x}\, dx =$$

$$= \frac{1}{\sqrt{2\pi}} \int_{-a}^{a} f(x)\, e^{-ip_1x+p_2x}\, dx$$

Since $\tilde{f}$ can be differentiated with respect to $p = p_1 + ip_2$, $\tilde{f}(p)$ is an analytic function. Since it has no poles it is entire. Furthermore, it is slowly increasing in the sense that

$$|p^n \tilde{f}(p)| \leq B_n\, e^{a|p_2|} \tag{1.11}$$

A theorem tells us that in fact every function that is infinitely differentiable and of bounded support $|x|<a$, has a Fourier transform which is entire analytic and satisfies (1.11) and vice versa. The space of all functions $\tilde{f}$ of this property is denoted by $\mathfrak{Z}$. The mapping $\mathcal{K}$ to $\mathfrak{Z}$ is one to one.

An analytic functional is a generalized function with test function space $\mathfrak{Z}$ and a contour Γ in the complex plane. Thus, if $\mathfrak{Z}'$ is the space of analytic functionals,

$$(T(\mathfrak{z}),\ f(\mathfrak{z})) = \int_\Gamma T^*(\mathfrak{z})\, f(\mathfrak{z})\, d\mathfrak{z}$$

for $T \in \mathfrak{Z}'$, $f \in \mathfrak{Z}$, and Γ a contour in the complex plane.

2. Analytic Functionals and the Class Δ_Γ^n

The Cauchy formula can be used to generalize the Dirac delta function to an analytic functional. One defines*

$$(\delta_D(z), f) = \oint \delta(\zeta - z)\, f(\zeta) d\zeta = \frac{1}{2\pi i} \oint \frac{f(\zeta) d\zeta}{\zeta - z} \tag{2.1}$$

so that

$$\delta(\zeta - z) = \frac{1}{2\pi i} \frac{1}{\zeta - z} \tag{2.2}$$

(There can be no confusion between this δ with complex argument and the usual $\delta(x)$ of (1.4)).

A related functional is obtained by choosing Γ along the real axis; by definition

$$(\delta_P(x), f) \equiv \frac{1}{2\pi i} \int_{-\infty}^{\infty} \frac{f(\xi) - f(x)}{\xi - x}\, d\xi \tag{2.3}$$

with obvious generalization for other infinite paths. Eq. (2.3) is one way of writing the Cauchy principle part of the integral over $f(\xi)/(\xi - x)$. It is interesting that it has a generalization to poles of higher order. One defines

$$R \int_{-\infty}^{\infty} \frac{f(\xi)\, d\xi}{(\xi - x)^{n+1}} \equiv \int_{-\infty}^{\infty} \frac{f(\xi) - \sum_{\nu=0}^{n} \frac{1}{\nu!} (\xi - x)^{\nu} f^{(\nu)}(x)}{(\xi - x)^{n+1}}\, d\xi \tag{2.4}$$

with the integration carried out symmetrically about the pole. Obviously (2.3) is the special case $n = 0$. Eq. (2.4) is called the "regularization" of the otherwise undefined integral over $f(\xi)/(\xi - x)^{n+1}$.

---

* Since $T(x)$ will now be a real quantity (apart from trivial factors of i) it will be convenient to omit the * in the definition (1.3) from here on.

It is easily verified that

$$\frac{d^n}{dx^n} R \frac{1}{\xi - x} = n! \, R \frac{1}{(\xi - x)^{n+1}} \tag{2.5}$$

although this equation is far from trivial.

Eqs. (2.1) and (2.3) can be combined linearly to produce other analytic functionals defined by other paths, such as the well-known contours C_R, C_A, C_{1R} etc. For example,

$$(\delta_R(x), f) = (\delta_P(x) - \frac{1}{2}\delta_D(x), f) = \frac{1}{2\pi i} \int_{C_R} \frac{f(\zeta)d}{\zeta - x} \tag{2.6}$$

Upon differentiation one finds symbolically

$$\left.\frac{1}{(\zeta - x)^{n+1}}\right|_{R,A} = R \frac{1}{(\zeta - x)^{n+1}} \mp \frac{(-1)^n}{n!} i\pi\delta^{(n)}(\zeta - x) \tag{2.7}$$

In relativistic quantum field theory we are interested in functionals of the invariant

$$p^2 + m^2 \equiv \vec{p}^{\,2} - p^{o2} + m^2 \equiv \omega^2 - p^{o2} \quad .$$

Since the contour will be in the <u>complex</u> p^o plane and we have a four-dimensional space, we must choose our test function space to be the direct product space

$$\mathcal{S}(R^3) \otimes \mathcal{Z}(R) \equiv \mathcal{Y}(R^4) \quad .$$

The partial fraction decomposition permits us to reduce the problem to the previous one-dimensional case; with $f \in \mathcal{Y}(R^4)$

$$\frac{1}{p^2 + m^2} = \frac{1}{2\omega} \left(\frac{1}{\omega - p^o} + \frac{1}{\omega + p^o}\right) \tag{2.8}$$

$$(\Delta_\Gamma, f) = \frac{1}{(2\pi)^4} \int_\Gamma \frac{f(p)d^4p}{p^2 + m^2} =$$

$$= \frac{1}{(2\pi)^4} \int \frac{d^3p}{2\omega} \int_\Gamma f(\vec{p},\, p^o)\, dp^o \left(\frac{1}{\omega - p^o} + \frac{1}{\omega + p^o}\right) \qquad (2.9)$$

Thus, one deals with an analytic functional as far as the p^o integration is concerned and the d^3p integration is the test function integration over a tempered distribution.

The generalization of (2.8) is

$$\frac{1}{(p^2+m^2)^{n+1}} =$$

$$= \sum_{\nu=o}^{n} \binom{n+\nu}{\nu} \frac{1}{(2\omega)^{n+\nu+1}} \left[\frac{1}{(\omega-p^o)^{n-\nu+1}} + \frac{1}{(\omega+p^o)^{n+\nu+1}}\right]$$

(2.10)

so that one can write for all integers $n \geq 0$

$$(\Delta_\Gamma^n,\, f) = \frac{(-1)^n}{(2\pi)^4} \int_\Gamma \frac{f(p)\, d^4p}{(p^2+m^2)^{n+1}} \qquad (2.11)$$

in terms of partial fractions*. Eqs. of the type (2.4) can then be applied. In this way one obtains for example

$$(\Delta_P^n,\, f) = \frac{(-1)^n}{(2\pi)^4} \int d^3p\; R \int_{-\infty}^{\infty} \frac{f(\vec{p},p^o)dp^o}{(p^2+m^2)^{n+1}} \qquad (2.12)$$

and

$$(\Delta_1^n,\, f) = \frac{(-1)^n}{(2\pi)^4} \int d^3p \int_{C_1} \frac{2\pi}{n!}\, \delta^{(n)}(p^2 + m^2)\, f(p)\, dp^o$$

(2.13)

The contour C_1 consists of one circle around the pole, $p^o = \omega$, (clockwise), and one circle around the negative pole, $p^o = -\omega$, (counter-clockwise); $\delta^{(n)}(p^2 + m^2)$ is

* Note that this definition differs by a factor $(-1)^n$ from the definition of reference 22.

the n-th derivative with respect to $(p^o)^2$ of the δ defined in (1.13). It can be reduced to $\delta^{(\nu)}(\omega - p^o)$ and $\delta^{(\nu)}(\omega + p^o)$ by means of

$$\delta^{(n)}(p^2 + m^2) =$$

$$= \sum_{\nu=o}^{n} \frac{1}{\nu!} \frac{(n+\nu)!}{(n-\nu)!} \frac{1}{(2\omega)^{n+\nu+1}} \left[\delta^{(n-\nu)}(p^o - \omega) + \right.$$

$$\left. + (-1)^{n-\nu} \delta^{(n-\nu)}(p^o + \omega) \right] \tag{2.14}$$

Since the various Δ^n_Γ are linearly related and can all be expressed in terms of $\Delta^n_{\ p}$ and $\Delta^n_{\ 1}$, the above relations determine Δ^n_Γ for all the standard contours:

$$C,\ C_R,\ C_A,\ C_{1R},\ C_{1A},\ C_{\pm} \quad .$$

If we write

$$(\Delta^n_\Gamma,\ f) = \frac{1}{(2\pi)^2} \int_\Gamma \tilde{\Delta}^n_\Gamma(p)\ f(p)\ d^4p \tag{2.15}$$

we can define the Fourier transform

$$\Delta^n_\Gamma(x) = \frac{1}{(2\pi)^2} \int_\Gamma \tilde{\Delta}^n_\Gamma(p)\ e^{ipx}\ d^4p = \frac{(-1)^n}{(2\pi)^4} \int_\Gamma \frac{e^{ipx}\ d^4p}{(p^2+m^2)^{n+1}} \tag{2.16}$$

according to (1.10). $\Delta^n_\Gamma(x)$ is now defined on

$$\mathcal{S}(R^3(\vec{x})) \times \mathcal{D}(R(x^o)) \quad .$$

By means of this Fourier transform it is easy to prove on the basis of (1.8) and (1.16) that

$$\Delta_\Gamma^n(x) = \frac{1}{n!}\left(\frac{d}{dm^2}\right)^n \Delta_\Gamma(x) \quad . \tag{2.17}$$

This permits one to obtain explicit expressions for $\Delta_\Gamma^n(x)$ very easily.

The $\Delta_\Gamma^n(x)$ satisfy the differential equations

$$K^{n+1}\ \Delta_\Gamma^n(x) = -\ \delta(x) \qquad \Gamma = C_{R,A,P,1R,1A} \tag{2.18}$$

$$K^{n+1}\ \Delta_\Gamma^n(x) = \ 0 \qquad \Gamma = C,\ C_1,\ C_\pm \tag{2.19}$$

where $K = \Box - m^2$. The contours Γ in (2.18) and (2.19) will sometimes be indicated by I and H respectively: Δ_I^n and Δ_H^n .

The special case $m = 0$ is obtained by a limiting process. One defines

$$D_\Gamma^n(x) = \lim_{m\to o} \Delta_\Gamma^n(x) \quad , \tag{2.20}$$

and one finds that this limit does not exist for all cases. It exists for $\Gamma = C,\ C_{P,R,A}$ and all $n \geq 0$; but it exists for $\Gamma = C_\pm,\ C_{1R,1A},\ C_1$, only in the case $n=0$. This can easiest be made plausible by the observation that a coalescence of the poles $p^o = \pm\omega$ in the case when $\vec{p} = 0$ and $m \to 0$ results in a pinching of the contours in all those cases which do not exist for general n.

The generalized functions $\Delta_\Gamma^n(x)$ and $D_\Gamma^n(x)$ have thus been defined as elements of $\mathcal{S}'(\vec{x}) \otimes \mathcal{D}'(x^o)$. It is however also possible to define them as $\varepsilon\ \mathcal{S}'(x)$ as four-dimensional tempered distributions. To this end one compares the definition of $\delta_D(x)\ \varepsilon\ \mathcal{S}'$ as in (1.4) with the path following the real line and $\delta_D(\mathfrak{z})\ \varepsilon\ \mathcal{Z}'$ as in (2.1) for $\mathfrak{z} = x$ real. Since $\delta_p^{(n)}(x)$ can clearly be defined either $\varepsilon\ \mathcal{Z}'$ or $\varepsilon\ \mathcal{S}'$, eq. (2.7) can be formally taken over with $\zeta = \xi$ real as a generalized function

$\varepsilon\ \mathcal{S}'$. This has the advantage that the Fourier transform, i.e. $\Delta^n_\Gamma(x)$, will also be defined $\varepsilon\ \mathcal{S}'(R^4)$ and thus has a larger test function space than $\mathcal{S}(R^3) \otimes \mathcal{D}(R)$.

While this is an advantage in certain respects it is a disadvantage in the study of the interrelations between the Δ^n_Γ for different Γ . As analytic functionals the topological nature of the paths relative to the two singularities in p-space specifies the linear combination of one path in terms of others. This feature is lost if one works with tempered distributions in all four variables.

Details of the subject matter discussed here can be found in reference 22 where also further references to distribution theory are given.

3. The Free Field Equation $K^{n+1}a(x) = 0$

The Cauchy initial value problem of this equation can be obtained in a rather standard way. We shall outline its derivation only briefly.

Consider the expression

$$b(x)(\square - m^2)^{n+1}\ a(x) - a(x)(\square - m^2)^{n+1}\ b(x) =$$

$$= b(x) \sum_{\nu=o}^{n+1} \binom{n+\nu}{\nu} \square^\nu (-m^2)^{n-\nu+1} a(x) - a(x)\Sigma \ldots b(x) \quad (3.1)$$

The term with $\nu=0$ cancels. By differentiation by parts one finds

$$b\,\square^{k+1}\, a - a\,\square^{k+1}\, b = \partial^\mu \sum_{\alpha=o}^{k} (\,\square^\alpha b(x) \overleftrightarrow{\partial}_\mu\, \square^{k-\alpha} a(x)) \quad (3.2)$$

where $\overleftrightarrow{\partial}_\mu$ means "∂_μ to the right minus ∂_μ to the left". With the aid of this equation one has, changing ν to $\alpha = \nu - 1$

$$b K^{n+1} a - a K^{n+1} b =$$

$$= \sum_{\alpha=0}^{n} \binom{n+1}{\alpha+1} (-m^2)^{n-\alpha} \partial^\mu \sum_{\beta=0}^{\alpha} (\Box^\beta b \overleftrightarrow{\partial}_\mu \Box^{\alpha-\beta} a) \qquad (3.3)$$

We now choose for b(x) the analytic functional Δ^n which we discussed in the previous lecture. Since it satisfies (2.19) we find, using Gauss'theorem,

$$0 = \int [\Delta^n(x-y) K_y^{n+1} a(y) - a(y) K_y^{n+1} \Delta^n(x-y)] \, d^4y =$$

$$= \Big(- \int_{\sigma \text{ through } x} + \int_{\substack{\sigma \text{ not} \\ \text{through } x}}\Big) \sum_{\alpha=0}^{n} \sum_{\beta=0}^{\alpha} \binom{n+1}{\alpha+1} (-m^2)^{n-\alpha} \times$$

$$\times \int \Box_y^\beta \Delta^n(x-y) \overleftrightarrow{\partial}_\mu^{\,y} \Box_y^{\alpha-\beta} a(y) \, d^3\sigma^\mu(y) \qquad (3.4)$$

One now observes that $\Delta^n(x)$ satisfies

$$\Delta^n(x)\Big|_{t=0} = 0 \qquad (x^0 = 0) \qquad (3.5a)$$

$$\left(\frac{\partial}{\partial t}\right)^\alpha \Delta^n(x)\Big|_{t=0} = 0 \qquad (1 \leq \alpha \leq 2n) \qquad (3.5b)$$

$$\left(\frac{\partial}{\partial t}\right)^{2n+1} \Delta^n(x)\Big|_{t=0} = (-1)^n \delta_3(\vec{x}) \qquad (3.5c)$$

where $\delta_3(\vec{x})$ is the three-dimensional Dirac delta function. These relations will be proved below. Using them in (3.4) results in

$$a(x) = \sum_{\alpha=0}^{n} \binom{n+1}{\alpha+1} (-m^2)^{n-\alpha} \sum_{\beta=0}^{\alpha} \int_{y^0=\text{const.}} \Box_y^\beta \Delta^n(x-y) \overleftrightarrow{\partial}_0^{\,y} \times$$

$$\times \Box_y^{\alpha-\beta} a(y) \, d^3y \qquad (3.6)$$

which is the desired solution of the Cauchy problem.

The proof of (3.5) is simplest in terms of the reduction relation (2.17)

$$\left(\frac{\partial}{\partial t}\right)^s \Delta^n(x) = \frac{1}{(2\pi)^3}\int e^{i\vec{p}\vec{x}}\, d^3p\, \frac{1}{n!}\left(\frac{d}{d\omega^2}\right)^n\left(\frac{\partial}{\partial t}\right)^s \frac{\sin\omega t}{\omega} \tag{3.7}$$

If s is even the time derivative gives

$$(-1)^{s/2}\,\frac{\sin\omega t}{\omega}\,\omega^s = (-1)^{s/2}\; t^{-s+1}\,\frac{\sin\omega t}{\omega t}(\omega t)^s$$

so that if one introduces $\alpha = \omega t$ instead of ω the integrand is an even regular function of α and an odd function of t and therefore vanishes for t = 0. This proves (3.5a) and (3.5b) for n even.

If s is odd the same procedure yields

$$\left(\frac{\partial}{\partial t}\right)^{2r+1} \Delta^n(x) = \frac{1}{(2\pi)^3}\int e^{i\vec{p}\vec{x}}\, d^3p\, \frac{1}{n!}\left(\frac{d}{d\omega^2}\right)^n(-1)^r\omega^{2r}\cos\omega t \tag{3.8}$$

$$\left(\frac{d}{d\omega^2}\right)^n\omega^{2r}\cos\omega t = t^{2(n-r)}\left(\frac{d}{d\alpha}\right)^n\alpha^r\cos\alpha\Big|_{\alpha=\omega t} =$$

$$=\begin{cases} 0 & \text{for } r < n \text{ and } t = 0 \\ n! & \text{for } r = n \text{ and } t = 0 \end{cases} \tag{3.9}$$

This completes the proof of (3.5b) and yields for r = n

$$\left(\frac{\partial}{\partial t}\right)^{2n+1}\Delta^n(x)\Big|_{t=0} = \frac{1}{(2\pi)^3}\int e^{i\vec{p}\vec{x}}(-1)^n d^3p = (-1)^n\delta_3(\vec{x}),$$

i.e. (3.5c).

We shall leave the physical interpretation of the equation $K^{n+1}a(x) = 0$ for later but note here that this classical field equation can be quantized in the following manner.

The commutator $[a(x), a(y)]$ must satisfy the following conditions:

(a) $K_x^{n+1}[a(x), a(y)] = 0$ because of the field equation

(b) $[a(x), a(y)] = 0$ for $(x-y)^2 > 0$ (micro-causality)

(c) $[a(x), a(y)] = - [a(y), a(x)]$,

(d) $(\frac{\partial}{\partial x^o})^{\alpha}[a(x), a(y)] = 0$ for $x^o = y^o$ and $0 \leq \alpha \leq 2n$.

This last condition is necessary in order to exclude solutions of lower order wave equations ($K^{\nu}a = 0, \nu < n+1$).

This together with Poincaré invariance determines the right side of the commutator up to a numerical factor if it is assumed to be a multiple of the unit operator; we choose the factor such that

$$[a(x), a(y)] = - i \Delta^n(x - y) \tag{3.10}$$

Clearly Δ^n is the only antisymmetric functional satisfying (2.19).

Finally, two remarks are in order. The solution of $K^{n+1}a(x) = 0$ can be expressed as a plane wave expansion involving creation and annihilation operators, exactly as in the special case n=o.

Secondly, the solution (3.6) can be used to define mass shell wave packets $f_{\alpha}(x) \in \mathcal{S}(R^3)$ as in the usual formulation.

4. Operator Derivatives

A differential calculus of derivatives of operators with respect to free field operators can be developed along the following lines.

We assume that we are dealing with operators that can be expressed in terms of the free fields a(x):

$$F = \sum_{n=o}^{\infty} \frac{(-i)^n}{n!} \int f(x_1 \ldots x_n) : a(x_1) \ldots a(x_n) : d^4x_1 \ldots d^4x_n \tag{4.1}$$

The symbol :...: denotes the usual Wick-ordering. The symmetry of the latter requires the symmetry of $f(x_1 \ldots$ $\ldots x_n)$.

Since the commutation relations of the a(x) are known the quantity $[a(x),F]$ can be computed. This commutator can be extended to an "inhomogeneous" commutator $[a(x),F]_I$ which is defined according to the solutions Δ_I^n of the inhomogeneous equation (2.18). The I refers to one of the subscripts R, A, P, 1R, 1A. For example, for $\Delta_R^n(x) = \Theta(x)\,\Delta^n(x)$ one defines $[a(x),F]_R$ by reducing $[a(x),F]$ down to the use of (2.10) and then replacing Δ^n by Δ_R^n.

The operator derivative associated with the free field a(x) satisfying

$$K^{n+1} a(x) = 0 \tag{4.2}$$

is now introduced by the definition

$$i \frac{\delta F}{\delta a(x)} \equiv K_x^{n+1} [a(x),F]_I \tag{4.3}$$

This definition is independent of the particular kind I of the generalized function Δ_I^n chosen.

For $F = \lambda 1$, where 1 is the unit operator,(3.2) obviously gives zero. For $F = a(y)$ one has

$$\frac{\delta a(y)}{\delta a(x)} = - i K_x^{n+1} [a(x), a(y)]_I = - i K_x^{n+1}(-i)\Delta_I^n(x-y) =$$

$$= + \delta(x-y) \qquad (4.4)$$

From this one obtains with (4.1)

$$i \frac{\delta F}{\delta a(x)} = \sum_{n=o}^{\infty} \frac{(-i)^n}{n!} \int f(x,x_1 \dots x_n) : a(x_1) \dots a(x_n) : \times$$

$$\times d^4x_1 \dots d^4x_n$$

Repeated applications of this derivative and the observation that the vacuum expectation value of a normal ordered product vanishes while $<0|:1:|0> = 1$, leads to

$$F = \sum_{n=o}^{\infty} \frac{1}{n!} <0| \frac{\delta^n F}{\delta a(x_1) \dots \delta a(x_n)} |0> : a(x_1) \dots a(x_n) : \times$$

$$\times d^4x_1 \dots d^4x_n \qquad (4.5)$$

This means that - apart from a phase factor - the "expansion coefficients" $f(x_1 \dots x_n)$ in (4.1) are just the vacuum expectation values of the n-th operator derivative. The analogy to the McLaurin series is apparent.

We note that the usual properties of derivatives are satisfied by (4.3):

$$\frac{\delta(\alpha F + \beta G)}{\delta a(x)} = \alpha \frac{\delta F}{\delta a(x)} + \beta \frac{\delta G}{\delta a(x)} \qquad (4.6)$$

$$\frac{\delta(FG)}{\delta a(x)} = \frac{\delta F}{\delta a(x)} G + F \frac{\delta G}{\delta a(x)} \qquad (4.7)$$

Also

$$\frac{\delta^2 F}{\delta a(x)\delta a(y)} = \frac{\delta^2 F}{\delta a(y)\delta a(x)} \tag{4.8}$$

While we have restricted ourselves here to the neutral scalar field, it is not difficult to extend these considerations to a charged field and to spin 1/2 and spin 1, which are the most important cases of our present interest.

For example, for spin 1/2 one has

$$-i\frac{\delta F}{\delta\bar{\psi}(x)} = (\gamma^\mu\partial_\mu + m)\ [\psi(x),\ F]_{\mp I} \tag{4.9}$$

$$-i\frac{\delta F}{\delta\psi(x)} = [F,\ \bar{\psi}(x)]_{\mp I}(-\gamma^\mu\overleftarrow{\partial}_\mu + m) \tag{4.10}$$

Here the $\pm$ alternative indicate commutator or anticommutator depending on whether F is a tensor or a spinor (i.e. a representation of the Lorentz group (m/2, n/2) with m + n even or odd).

In this connection one should compare the operator derivative with the functional derivative often used in field theory. The latter leads to the mathematically meaningless concept of "anti-commuting c-numbers"; the operator derivative involves no such difficulties.

Another way of looking at the operator derivative is in terms of p-space. But we shall not make use of it here so that I shall restrict myself only to the basic defining equation for the simple case n = 0:

$$i\frac{\delta F}{\delta a(p)} \equiv -(p^2 + m^2)[a_m(p),\ F]_I \tag{4.11}$$

Here $a_m(p)$ is the (four-dimensional) Fourier transform of a(x).

An important concept is that of <u>strong</u> and <u>weak</u> equations. A strong operator equation is one which can be operator differentiated any number of times and remains valid. Otherwise, the equation is called weak. Obviously, operator differentiation will in general not leave an operator equality correct.

The simplest example is provided by the free field equation when it is assumed that the operator derivative and the ordinary derivative are to commute:

$$\frac{\delta}{\delta a(x)} K_y^{n+1} a(y) = K_y^{n+1} \frac{\delta}{\delta a(x)} a(y) = K_y^{n+1} \delta(x-y) \quad (4.12)$$

Since this quantity does not vanish, $K_y^{n+1} a(y) = 0$ does <u>not</u> imply

$$\frac{\delta}{\delta a(x)} K_y^{n+1} a(y) = 0$$

so that (3.2) is a <u>weak</u> equation. (Until fairly recently the published literature assumed (4.2) to be a strong equation in which case the operator derivative and the ordinary derivative cannot commute. This leads to rather unpleasant technical complications, so that this assumption has now been generally abandoned).

An example of a strong equation is the commutation relation (3.10). The operator derivative on the left leads to the commutator of a(x) or a(y) with $\delta(x-y)1$ which clearly vanishes, while on the right is the operator derivative of a multiple of the unit operator which vanishes too.

We summarize these two important results by writing

$$K^{n+1} a(x) \overset{W}{=} 0 \quad (4.13)$$

and

$$[a(x), a(y)] \overset{S}{=} - i \, \Delta^n (x - y) \quad (4.14)$$

The calculus of operator derivatives was developed

in ref. 8 and was later extended to p-space in reference 16. The assumption that the operator derivative and the ordinary derivative commute and that therefore the free field equation must be considered as weak was first made by Chen (ref. 19).

5. The Operators P^N_{12} and their Algebra

The importance of the step function $\Theta_{12} \equiv \Theta(x_1 - x_2)$ is well known, and so are the difficulties to which it gives rise since even the product with the simplest tempered distribution,

$$\delta_{12} \equiv \delta(x_1 - x_2) \quad ,$$

i.e. $\Theta_{12}\delta_{12}$ is not defined (cf. (1.7) and the remarks following it). The operators P^N_{12} to be defined below can be regarded as generalizations of Θ_{12}, which can multiply δ_{12} and its derivatives of higher and higher order as N increases.

We define the integral operator P^N_{12} (N > 0 integer) by

$$P^N_{12} f(x_1, x_2) \equiv$$

$$\equiv (K_1 K_2)^N \Theta_{12} \int \Delta^{N-1}_A (x_1 - \xi) \Delta^{N-1}_R (x_2 - \eta) f(\xi, \eta) d\xi d\eta \qquad (5.1)$$

For N = 0 one can define

$$P^0_{12} f(x_1, x_2) = \Theta_{12} \int \delta(x_1 - \xi) \delta(x_2 - \eta) f(\xi, \eta) d\xi d\eta =$$

$$= \Theta_{12} f(x_1, x_2) \qquad (5.2)$$

P^N_{12} is thus an integral operator in terms of generalized functions. In the following it will be convenient to regard $\Delta^N_\Gamma \in \mathcal{S}'$ and the convolution product (5.1) will exist for $f(\xi,\eta) \in \mathcal{S}'$ if it is suitably restricted.

Let $\mathcal{S}'_{4N}(x_1^o,x_2^o)$ be the set of tempered distributions of two variables which can be expressed in the form

$$\sum_{mn} a(x_1,x_2)\left(\frac{\partial}{\partial x_1^o}\right)^m \left(\frac{\partial}{\partial x_2^o}\right)^n \delta(x_1^o - x_2^o) \tag{5.3}$$

$(m + n < 4N)$

Then one can show that $P_{12}^N f(x_1,x_2)$ is defined for all $f \in \mathcal{S}'_{4N}$.

The operators P_{12}^N have the following important properties:

$$P_{12}^N P_{12}^N = P_{12}^N \tag{5.4}$$

$$P_{12}^N P_{21}^N = 0 \tag{5.5}$$

One can form products of the type

$$P_{1\ldots m}^N \equiv P_{12}^N P_{23}^N \ldots P_{m-1\,m}^N \tag{5.6}$$

These products are always defined since they are convolution products with one factor of compact support. Let $\mathcal{P}_m'^N$ be the linear space of functionals of the type (5.6) involving m fourvectors x_k. Clearly, $\mathcal{P}_m'^N \subset \mathcal{S}'$. Let $\mathcal{P}_m$ be the space of functions $\in \mathcal{S}$ in m fourvectors x_k; $\mathcal{P}_m \subset \mathcal{S}$. Then we define a space $\mathcal{F}'_m$ of tempered distributions by

$$\mathcal{F}_m'^N = \{f_m P_{1\ldots m}^N;\ f_m \in \mathcal{P}_m,\ P_{1\ldots m}^N \in \mathcal{P}_m'^N\} \tag{5.7}$$

Thus $\mathcal{F}_m'^N$ is essentially the space obtained from the $P_{1\ldots m}^N$ by smearing from the left. Since $P_{1\ldots m}^N$ depends on 2m variables $f_m P_{1\ldots m}^N$ is still a distribution in m variables.

One can now prove that on $\mathcal{F}'^N_m$ the P^N_{ij} form an Abelian algebra,

$$[P^N_{ij}\ ,\ P^N_{kl}] = 0 \qquad (5.8)$$

with ij, kl not necessarily different. This very basic result permits one to derive many important consequences. In particular, if one defines

$$B^N_{12} \equiv \mathbf{1}_{12} - P^N_{12} - P^N_{21} \qquad (5.9)$$

where

$$\mathbf{1}_{12} \equiv \delta(x_1 - \xi)\,\delta(x_2 - \eta)\ , \qquad (5.10)$$

then

$$B^N_{12} B^N_{12} = B^N_{12} \qquad \text{and} \qquad B^N_{12} P^N_{12} = B^N_{12} P^N_{21} = 0 \qquad (5.11)$$

and one can prove that (using N = 1)

$$\begin{aligned} B_{1\ldots m} &\equiv B_{1\ldots m}(x_1 \ldots x_m;\ \xi \ldots \omega) = \\ &= -\,K_1 \ldots K_m \sum_{k=1}^{m} \Theta^x_{k1} \ldots \Theta^x_{kk-1}\ \Theta^x_{kk+1} \ldots \Theta^x_{km}\ \times \\ &\quad \times\ \Delta_R(x_k - \xi_k) \prod_{\substack{i=1 \\ i \neq k}}^{m} \Delta(x_i - \xi_i) \end{aligned} \qquad (5.12)$$

satisfies the equation

$$B_{1\ldots m} = \sum_{i>j=1}^{m} B_{ij} \qquad (5.13)$$

on $\mathcal{F}'_m$. This result will be used in the current formulation (Sections 8 and 9).

Let $F(x_1, x_2) \in \mathcal{S}'_{4N}$ and let

$$\text{supp } F(x_1, x_2) = \bar{V}_+(x_2 - x_1) = \bar{V}_-(x_1 - x_2) \qquad (5.14)$$

then

$$P^N_{12} F(x_2, x_1) + P^N_{21} F(x_1, x_2) = 0 \quad . \qquad (5.15)$$

This relation will be important in connection with causality which is a statement of the form (5.14). Its proof is related to (5.5).

The Abelian algebra of the P_{ij} on $\mathcal{F}'$ was developed by Wray (ref. 23).

B. Asymptotic Quantum Field Theory

The development of this theory is a natural extension of the ideas of Lehmann, Symanzik and Zimmermann, of Bogoljubov and his collaborators, and of Nishijima and his students. It has presently reached a plateau which can perhaps best be characterized by the word "perturbation expansion". What has been accomplished is to formulate a theory free of divergences and related ambiguities which besets the standard field theory of the late fourties and early fifties. At the same time, it was not possible so far to solve the basic equations in other ways than by perturbation expansion. While this expansion probably does not converge in general, each term of it is finite and free of divergences. Nor does it require renormalization since only physical quantities enter the theory. Suitable boundary conditions are specified which make the scattering matrix unique.

However, the results of perturbation expansion will be seen to be physically meaningful only for those interactions which correspond to the so-called "renormalizable" types. For the "unrenormalizable" case finite results can also be obtained in any given order of the expansion, but this expansion does not seem to be physically meaningful, because additional constants arise in each additional order. The present development which will be summarized in the following lectures must therefore be regarded as only preliminary for the nonrenormalizable case, but satisfactory within the applicability of perturbation expansions for the renormalizable case.

In order to demonstrate the essential features of this theory it will suffice to discuss the self-interacting neutral scalar field. More general cases will be discussed in the last lecture.

6. Causality and the GLZ Expansion

The fundamental assumptions of asymptotic quantum field theory are

1. The quantum field theory of the relativistic free field.

This theory is fully understood mathematically, especially for the spin zero case which we shall use explicitly, $Ka(x) = 0$. However we shall work with the generalization

$$K^N a(x) \overset{W}{=} 0 \qquad (6.1)$$

which was discussed in Section 3, where the relevant commutation relation was also given. The usual assumptions on the vacuum, the spectrum, and on spin-statistic relations are part of this free field theory.

2. Strong unitarity of the S-operator.

We assume completeness of the asymptotic fields $a_{in}(x)$ and of the asymptotic fields $a_{out}(x)$. The Hilbert spaces $\mathcal{H}_{in}$ and $\mathcal{H}_{out}$ spanned by their polynomials upon completion are unitarily equivalent. Thus a unitary operator S such that

$$a_{out}(x) \overset{S}{=} S^{*} a_{in}(x) S \qquad (6.2)$$

exists. We assume further that it is <u>strongly</u> unitary.

$$S^{*} \overset{S}{=} S^{-1} \qquad (6.2')$$

3. Causality

The usual micro-causality assumption (locality)*

$$[a(x), a(y)] = 0 \qquad x \sim y \qquad (6.3)$$

is replaced by a stronger assumption first suggested by Bogoljubov. This suggestion is modified by using operator derivatives and by requiring a strong equation. Defining**

$$J(x) = i S^{*} \frac{\delta S}{\delta a(x)} \quad , \qquad (6.4)$$

the causality statement is***

$$\frac{\delta J(x)}{\delta a(y)} = 0 \qquad (x \overset{<}{\sim} y) \quad . \qquad (6.5)$$

It can also be expressed by

$$\text{supp} \frac{\delta J(x)}{\delta a(y)} = \bar{V}_{+}(x-y) \qquad (6.6)$$

where $V_{+}(x)$ is the future light cone of the vector x, i.e. the vector x can point only into the future.

At this point one can proceed in two different ways,

---

* The symbol $x \sim y$ means x and y are spacelike to each other and x = y is excluded.

** From here on we shall always work with a_{in} and shall therefore omit the subscript "in".

*** $x < y$ means x and y are timelike to each other and $x^{o} < y^{o}$.

depending on whether or not one wants to make use of an interpolating field A(x) (which interpolates between t = - ∞ and t = + ∞). If one does, one develops the field formalism. Otherwise one works with the current formalism. The latter can also be obtained from the former by elimination of A(x) but it is not necessary to proceed that way. We shall discuss the current formalism in Sections 8 and 9. The present Section as well as Section 7 will be based on the field formalism, although various equations will remain useful for the later development.

An interpolating field can be defined by means of an integral equation such as

$$A(x) \stackrel{S}{=} a(x) - \int \Delta_R^n(x - y)\, J(y)\, d^4y \quad , \tag{6.7}$$

or equivalently in terms of the more compact expression

$$A(x) \stackrel{S}{=} S^*(a(x)S)_+ \quad . \tag{6.8}$$

The subscript + indicates positive time ordering: one uses Wick's theorem to arrange the expansion (4.1) for the S-operator,

$$S \equiv \mathbf{1} + \sum_{n=1}^{\infty} \frac{(-i)^n}{n!} \int \omega(x_1 \ldots x_n) \, :a(x_1) \ldots a(x_n): \times \\ \times\, d^4x_1 \ldots d^4x_n \tag{6.9}$$

in terms of time-ordered products and then one places a(x) at the appropriate position.

It is not difficult to see that (6.7) and (6.8) are equivalent since the ordering theorem

$$(a(x)F)_+ \stackrel{S}{=} F\, a(x) + [a(x),F]_R \tag{6.10}$$

permits an easy proof. In any case, the interpolating

field can be defined as a field which approaches a_{in} (a_{out}) in the asymptotic limit $t \to -\infty$ ($t \to +\infty$). Thus one must satisfy the asymptotic condition:

$$\lim_{\sigma \to \mp\infty} \langle\Psi| \int_\sigma \Delta^n(x-y)\overleftrightarrow{\partial}^y_\mu A(y)d^3\sigma^\mu(y)|\Phi\rangle = \langle\Psi|a_{\substack{in\\out}}(x)|\Phi\rangle \tag{6.11}$$

with $|\Psi\rangle$, $|\Phi\rangle \in \mathcal{H}$. While this limit is in the space of generalized functions, it is simpler to regard this as a weak limit in $\mathcal{H}$ by "smearing" both sides first over the appropriate test function space.

It is to be noted that this asymptotic condition is part of the definition of the interpolating field. It is not one of the basic assumptions of the theory since it is apparently not needed if one works with the current formalism.

An operator equation from which A(x) can be computed (at least in principle)is the expansion first suggested by Glaser, Lehmann, and Zimmermann (ref. 3). The following generalization of this equation can be proven:

$$A(x) \overset{s}{\equiv} a(x) + \sum_{n=2}^{\infty} \frac{1}{n!} \int K_1^N \ldots K_n^N \langle R(x;x_1 \ldots x_n)\rangle_o \times$$
$$\times :a_1 \ldots a_n: d^4x_1 \ldots d^4x_n \tag{6.12}$$

where, using $\Theta_{1\ldots n} \equiv \Theta_{12}\Theta_{23}\ldots\Theta_{n-1n}$ and $A_k \equiv A(x_k)$,

$$R(x;x_1 \ldots x_n) \equiv (-i)^n \sum_{perm} \Theta_{x1\ldots n} \times$$
$$\times [\ldots[[A_x,A_1],A_2],\ldots A_n] \quad (n>0) \tag{6.13}$$

is the "retarded product".

More generally, in fact

$$\frac{\delta^n A(x)}{\delta a(x_1)\ldots\delta a(x_n)} \stackrel{s}{=} K_1^N \ldots K_n^N \; R(x;x_1\ldots x_n) \qquad (n > 0)$$

(6.14)

This differs from ref. 3 in four respects. First, it is an operator equation for the n-th derivative of A(x) and does not only involve the vacuum expectation value of (6.14), secondly it involves operator derivatives, thirdly these are strong equations, and fourthly the basic free field is a solution of (6.1) with arbitrarily (but fixed) N and is not only valid for N=1.

The proof of (6.14) is not trivial and rather long. It proceeds in three steps. One proves it first for n=1, then one proves the lemma

$$\frac{\delta}{\delta a(x_{n+1})} R(x;x_1\ldots x_n) \stackrel{s}{=} K_{n+1}^N \; R(x;x_1\ldots x_{n+1}) \;,$$

(6.15)

and finally one completes the proof by induction. We must refer to the literature for details (refs. 18 and 19).

Equations (6.12) and (6.14) can be combined to yield for $n > 0$

$$K_1^N \ldots K_n^N \; R(x;x_1\ldots x_n) =$$

$$= \sum_{m=o}^{\infty} \frac{1}{m!} \int K_1^N \ldots K_{n+m}^N \langle R(x;x_1\ldots x_n x_{n+1}\ldots x_{n+m})\rangle : a_{n+1}\ldots$$

$$\ldots a_{n+m} : d^4x_{n+1}\ldots d^4x_{n+m}$$

(6.16)

This equation or (6.15) would have to be solved for $R(x;x_1\ldots x_n)$ in order to determine A(x) and its derivatives. This has been carried out so far only in perturbation expansion. A non-perturbative solution seems to be very difficult to obtain.

In any case the GLZ expansion does not appear to be a fruitful approach towards the dynamics of particles since the solution for A(x) is at least as difficult and in any case less direct than the equations for the current or the S-operator which we shall consider next. References 1, 4, 6, 17 for causality.

7. The Pugh Equation

The integral equation (6.7) gives A(x) as a functional of the S-operator. We shall now effectively invert this equation, obtaining the S-operator as a functional of A(x). We shall restrict ourselves to the case N=1 in the present Section.

We define

$$J_n(x_1 \ldots x_n) \equiv i^n S^* \frac{\delta^n S}{\delta x_1 \ldots \delta x_n} \qquad (7.1)$$

Clearly $J_1(x) \equiv J(x)$. Next we define the ϕ-product of the interpolating field. Given the relationship between the normal ordered and the time-ordered products of free fields (Wick's theorem), the ϕ-product has exactly the same relationship to the time-ordered product when one is dealing with A(x) instead of a(x). The inversion of (6.7) is then found to be (refs. 18 and 23),

$$J_n(x_1 \ldots x_n) \overset{W}{=} K_1 \ldots K_n \; \phi(A(x_1) \ldots A(x_n)). \qquad (7.2)$$

For n=1 this relation follows from the strong equality

$$J(x) \overset{S}{=} K\, A(x) - K\, a(x) \qquad (7.3)$$

obtained from (6.7) by operation with K. For n > 1 it follows by induction.

A somewhat lengthy derivation now leads from (7.2) to the Pugh equation. One can proceed either via operator equations (M. Wilner, unpublished) or via matrix elements (Pugh) to derive the important equation (Pugh equation, refs. 7 and 23).

$$(1 - B_{1\ldots m})\, \omega(x_1 \ldots x_m) = \lambda(x_1 \ldots x_m) \tag{7.4}$$

Here $B_{1\ldots m}$ is defined in (5.12), ω are the S-matrix elements of (6.9) and are related to J_n by

$$\omega(x_1 \ldots x_m) = \langle J_m(x_1 \ldots x_m)\rangle_o \quad , \tag{7.5}$$

and λ is the vacuum expectation value of the operator valued tempered distribution

$$K_1 \ldots K_m \Big[\sum_{K=2}^{m} \phi(\alpha_1 \ldots \alpha_K,\, a_{K+1} \ldots a_m) + {} + \phi((1-S)\alpha_1 a_2 \ldots a_m)\Big] \tag{7.6}$$

with $\alpha(x)$ being the second term on the right of (6.7).

The integral operator $B_{1\ldots m}$ is idempotent and λ satisfies $B\lambda = 0$ so that a formal solution of (7.4) can be obtained in the form

$$\omega(x_1 \ldots x_m) = \chi(x_1 \ldots x_m) + \lambda(x_1 \ldots x_m) \quad . \tag{7.7}$$

The quantity χ is a solution of the homogeneous equation

$$(1 - B_{1\ldots m})\, \chi(x_1 \ldots x_m) = 0 \tag{7.8}$$

The most general solution of this equation is any tempered distribution for which the convolution product in (7.8) is defined and which is $\varepsilon\, \mathcal{S}'_4$ (see (5.3)) with respect to any two variables x_i^o, x_j^o .

Of course, even if χ is known, eq. (7.7) is not a solution for ω since λ is not known. However, the great value of (7.7) lies in the fact that it does represent a solution for ω in perturbation expansion. In that case one finds

$$\omega^{(n)}(x_1,x_2\ldots x_m) = \chi^{(n)}(x_1\ldots x_m) + \lambda^{(n)}(x_1\ldots x_m) \tag{7.9}$$

for the n-th order term. This is in fact a recursive solution since $\lambda^{(n)}$ depends only on the $\omega^{(k)}$ with $k < n$. Thus, if $\lambda^{(1)}$ is chosen and the $\chi^{(n)}(n \geq 1)$ are fixed by the boundary conditions (see Section 9), we have indeed a unique solution of the S-matrix in any order of perturbation expansion.

Explicite calculations were actually carried out for quantum electrodynamics in lowest order. The results are identical with renormalized Feynman-Dyson theory. It is to be noted that no "bare" particles or fields are ever introduced and there is consequently no renormalization to be made in this theory.

In perturbation expansion it is clear that the choice of $\lambda^{(1)}$ essentially determines the interaction in this non-Lagrangian, non-canonical theory. We shall see later, however, that the boundary conditions also play an important role.

In a non-perturbative solution of (7.7) the choice of χ must be compatible with interaction and boundary conditions. Solutions will not exist for arbitrary choices of interaction.

8. The Current Formalism

As indicated earlier, it is possible to compute the S-matrix without ever introducing an interpolating field A(x). The first step in this direction is the derivation of the current equation (refs. 15, 17 and 23).

The strong unitarity assumption leads via operator differentiation and use of (6.4) immediately to

$$[J(x_1), J(x_2)] \stackrel{\text{s}}{=} i\left(\frac{\delta J(x_1)}{\delta x_2} - \frac{\delta J(x_2)}{\delta x_1}\right) \quad . \qquad (8.1)$$

Here we used the convenient notation $\frac{\delta}{\delta x} \equiv \frac{\delta}{\delta a(x)}$.
The integral operator (5.9) then leads to

$$(1 - B^N_{12})\, i\frac{\delta J_1}{\delta x} \stackrel{\text{s}}{=} P^N_{12}[J_1, J_2] + P^N_{12}\, i\,\frac{\delta J_2}{\delta x_1} + P^N_{21}\, i\frac{\delta J_1}{\delta x_2}$$

The causality assumption (6.5) together with (5.15) now tells us that the last two terms of this equation vanish, so that one obtains the <u>current equation</u>

$$(1 - B^N_{12})\, i\,\frac{\delta J_1}{\delta x_2} \stackrel{\text{s}}{=} P^N_{12}[J_1, J_2] \quad . \qquad (8.2)$$

Similar to the case of the Pugh equation a formal solution of this equation can be obtained because the "inhomogeneity" is orthogonal to B^N_{12} (see 5.11)). Thus,

$$i\,\frac{\delta J_1}{\delta x_2} = \chi(x_1, x_2) + P^N_{12}[J_1, J_2] \qquad (8.3)$$

with χ again a solution of the homogeneous equation

$$(1 - B^N_{12})\, \chi(x_1, x_2) = 0 \quad . \qquad (8.4)$$

The interaction in question is characterized by the choice of χ and affects also the boundary conditions that need to be satisfied.

If (8.2) or equivalently (8.3) is solved the S-operator follows easily by inverting (6.4),

$$i\,\frac{\delta S}{\delta x} = S\, J(x) \qquad (8.5)$$

which has the recursive solution (see (7.1)),

$$J_n(x_1 \ldots x_n) = J(x_n) J_{n-1}(x_1 \ldots x_{n-1}) + i\,\frac{\delta J_{n-1}(x_1 \ldots x_{n-1})}{\delta x_n} \qquad (8.6)$$

Alternatively, one can convert the current equation into an equation for the S-operator. Substitution of (6.4) and use of (5.9) yields

$$S^{*} \frac{\delta^{2} S}{\delta x_{1} \delta x_{2}} \stackrel{\mathrm{S}}{=} \chi'(x_{1}, x_{2}) - P^{N}_{12} \frac{\delta S^{*}}{\delta x_{1}} \frac{\delta S}{\delta x_{2}} - P^{N}_{21} \frac{\delta S^{*}}{\delta x_{2}} \frac{\delta S}{\delta x_{1}} \quad . \tag{8.7}$$

A term $B^{N}_{12} \frac{\delta S^{*}}{\delta x_{2}} \frac{\delta S}{\delta x_{1}}$ has here been combined with χ to give χ' , since this term is also $\varepsilon \mathcal{S}'_{4N}$.

Finally, one can obtain an equation very similar to the Pugh equation from the generalized current equation (ref. 23).

This is simply the generalization of (8.2) to higher derivatives of the current. A somewhat lengthy proof yields

$$\prod_{i>j=o}^{n} (1 - B^{N}_{ij}) \frac{\delta^{n} J(x)}{\delta x \ldots \delta x_{n}} \stackrel{\mathrm{S}}{=} R^{N}_{P}(x; x_{1} \ldots x_{n}) \tag{8.8}$$

with the R_P-product defined analogously to the R-product of equation (6.13),

$$R_{P}(x; x_{1} \ldots x_{n}) \equiv (-i)^{n} \sum_{\substack{\text{perm} \\ 1..n}} P^{N}_{x1\ldots n} [\ldots[[J_{x}, J_{1}], J_{2}], \ldots J_{n}] \tag{8.9}$$

and n > o. The similarity with (6.14) is apparent. Eq. (8.2) is clearly the special case n=2 of this results.

The S-operator equation (8.7) can be similarly generalized with the result

$$\prod_{i>j=o}^{n} (1 - B^{N}_{ij}) \, S^{*} \frac{i^{n} \delta^{n} S}{\delta x_{1} \ldots \delta x_{n}} \stackrel{\mathrm{S}}{=} P^{N}_{+}(J_{1} \ldots J_{n}) \tag{8.10}$$

with

$$P^{N}_{+}(J_{1} \ldots J_{n}) \equiv \sum_{\substack{\text{perm} \\ 1..n}} P^{N}_{1\ldots n} J_{1} \ldots J_{n} \tag{8.11}$$

Eq. (8.7), when operated on by $1 - B^N_{12}$ is clearly the special case n=2 of the general equation (8.10).

If one takes vacuum expectation values and observes (7.1) and (7.5) one has from (8.10) the S-matrix equation

$$\prod_{i>j=o}^{n} (1 - B^N_{ij})\, \omega(x_1 \ldots x_n) = \mu(x_1 \ldots x_n) \qquad (8.12)$$

with

$$\mu(x_1 \ldots x_n) = \langle P^N_+(J_1 \ldots J_n)\rangle_o \quad . \qquad (8.13)$$

The similarity with (7.4) is obvious.

We must remember that all these equations are valid on the space $\mathcal{F}'^N$ to the left, since only there do we have the abelian algebra of the P^N_{ij} which is used in these proofs.

This abundance of alternative equations for the computation of the S-matrix characterizes the fact that we know almost nothing about non-perturbative solutions to any of these. Perturbation calculations can of course be carried out relatively easily with each of them. Here, however, it is essential to be aware of the <u>type</u> of the theory one is dealing with.

The <u>type</u> of the theory is given by the exponent N which governs the free field equation (6.1). It is at our disposal and must be chosen in such a way that all expressions in the computations to any given order n of perturbation expansion are well defined.

We recall the need for choosing χ in (8.3) or χ' in (8.7) so that a particular interaction is described and the boundary conditions are satisfied. If $\chi \in \mathcal{S}'_{4N'}$ is necessary for this purpose then the type of the theory must be $N \geq N'$. If in particular the minimum allowed choice of the type is N=1, we call the theory "renormalizable" because it will then correspond to a renormalizable case of standard field theory. The essential point here is that this choice N=1 can be made no matter how high the order of expansion which one wants to calculate.

Theories of type N > 1 are called "nonrenormalizable". Here the situation is different. One can prove that, as one goes to higher orders in the expansion, a higher minimum choice of N is required. We shall return to this point in the following Sections.

9. Feynman Diagrams and Boundary Conditions

Consider the equation (8.12). Its "solution" is

$$\omega(x_1 \ldots x_n) = \chi(x_1 \ldots x_n) + \mu(x_1 \ldots x_n) \tag{9.1}$$

with χ a solution of the homogeneous equation, i.e. $\varepsilon\ \mathcal{S}'_{4N}(x_i^o, x_j^o)$ for every pair of variables. Now it can easily be shown that $B^N_{ij} = 1$ on the mass shell. This means that μ vanishes on the mass shell according to (8.2) and that therefore the ω which occur in the S-operator expansion (6.9) are given by

$$\omega(x_1 \ldots x_n)\Big|_{\text{mass shell}} = \chi(x_1 \ldots x_n) \quad . \tag{9.2}$$

But in order to determine χ we must use the boundary conditions and in order to specify the boundary conditions we must give physical meaning to the individual terms in ω by means of diagrams of the Feynman type. To this end we study μ as given by (8.13) and note that apart from c-number factors it is the vacuum expectation value of a product of currents. Equivalently, the latter are a product of operator derivatives of S or S^*; e.g.

$$J_1 J_2 = iS^* \frac{\delta S}{\delta x_1}\ iS^* \frac{\delta S}{\delta x_2} = + \frac{\delta S^*}{\delta x_1}\ SS^* \frac{\delta S}{\delta x_2} = \frac{\delta S^*}{\delta x_1}\frac{\delta S}{\delta x_2} \ .$$

These in turn are sums of products of normal ordered free fields,

$$:a(x_1)\ldots a(x_r):\ :a(y_1)\ldots a(y_s):$$

whose vacuum expectation values lead to products of Δ_+ and Δ_- functions. Each particular convolution product can then be associated with a diagram, though there will be in general several terms corresponding to the same diagram. Each line in such a diagram is no longer associated with the same "propagator" but can rather be either Δ_+ or Δ_- and there will be other terms arising from the P_+^N ordering. But we have

$$\mu = \Sigma\, \mu_{\text{Diagram}} \tag{9.3}$$

and therefore via (8.12)

$$\omega = \Sigma\, \omega_{\text{Diagram}} \tag{9.4}$$

each ω_{Diagram} being unambiguously associated with a diagram of a certain topology.

The advantage of the diagrammatic interpretation of individual terms of the scattering matrix elements lies in their physical interpretation. One can make the distribution between reducible and irreducible diagrams just as in Feynman-Dyson theory and one can thus give physical motivation to the boundary conditions.

The boundary conditions are best explained in terms of the example of the self-interacting neutral scalar field which we have been using all along. In addition to the obvious requirements of Poincaré invariance and symmetry with respect to permutations of its arguments the S-matrix elements $\omega(x_1 \ldots x_n)$ of (6.9) must also satisfy the following boundary conditions.

(1) For the two-point function

$$\lim_{p^2+m^2\to 0} \frac{\tilde{\omega}(p)}{(p^2+m^2)^2} \quad \text{exists}, \quad \neq 0 \tag{9.5}$$

where m is the observed (physical) mass . This condition is actually ensured by the stability of the vacuum ($S|0\rangle = |0\rangle$) and of the single particle states ($a_{in}(x)|0\rangle = a_{out}(x)|0\rangle$) . In perturbation expansion Eq. (9.5) must be (and is) satisfied for every order n.

(2) Corresponding to a $\phi^M(x)$ interaction of the customary (Lagrangian) formulation of field theory one requires

$$\tilde{\omega}_{\{p_i\to\alpha_i\}}(p_1\ldots p_M) = g\delta(\Sigma p) \quad , \tag{9.6}$$

which defines the coupling constant. The values α_i are usually chosen to be mass shell values whenever possible and zero otherwise, e.g.

$$\tilde{\omega}_{\substack{p_1\to 0\\ p_2^2+m^2\to 0\\ p_3^2+m^2\to 0}}(p_1p_2p_3) = g\,\delta(\Sigma p) \quad . \tag{9.7}$$

In perturbation expansion this condition becomes

$$\tilde{\omega}^{(1)}_{\{p_i\to\alpha_i\}}(p_1\ldots p_M) = g\,\delta(\Sigma p)\ ;$$

$$\tilde{\omega}^{(n)}_{\{p_i\to\alpha_i\}}(p_1\ldots p_M) = 0, \qquad n > 1 \tag{9.8}$$

In addition, it is also necessary to specify the (time-like) "high energy" behaviour of these "vertex" functions. For example,

$$\lim_{p_1^2\to-\infty} \frac{\tilde{\omega}^{(n)}(p_1p_2p_3)}{\sqrt{|p_1^2|}} \rightarrow 0 \qquad \text{all n.} \tag{9.9}$$

(3) All other m-point functions must vanish in the high energy limit:

$$\lim_{p_1^2 \to -\infty} \tilde{\omega}(p_1 \ldots p_m) = 0 \quad , \quad m > M \quad . \tag{9.10}$$

The work discussed in this section is largely due to Dr. Wray (ref. 23). See also ref. 7, where some boundary conditions are however not correctly stated.

10. Summary

These lectures presented very briefly the present state of asymptotic quantum field theory. Actual applications of the theory had to be omitted, however, due to lack of time available. They would also be rather boring in a lecture series. It is apparent from a comparison with the previous such summary (ref.20) how much progress was made in less than two years. A comparison with conventional quantum field theory of the late forties and early fifties is also worthwhile. Perhaps the salient points of such a comparison are the following.

Mathematically, the theory of generalized functions is now playing a very essential role and helps greatly in establishing a mathematically satisfactory theory (from the theoretical physicist's point of view) which is free of divergent and ill-defined expressions. The structure of the newly emerging theory is also very different: it is <u>not</u> a Lagrangian theory and it does <u>not</u> involve a canonical formulation. Furthermore, it does not contain the notion of bare particle or that of renormalization. The boundary conditions ensure that no other but physical quantities are introduced (specifically masses and coupling constats). There are therefore also no renormalization constants appearing.

In discussing the results of the theory a distinction is desirable between the theories of type 1 ($N = 1$, $Ka = 0$) and those of type $N > 1$($K^N a = 0$).

Theories of type 1 correspond to the renormalizable interactions of conventional quantum field theory. This is apparent when one notices the restriction of $\chi(x,y)$ to $\varepsilon\ \mathcal{S}_4'$. For this type the boundary conditions can be satisfied in each order of perturbation expansion by introducing a finite (small) number of constants (masses, coupling constants). This number need not be increased as one goes to higher orders. Specifically, the Pugh equation was solved in perturbation expansion for the case of quantum electrodynamics of spin 1/2. It was shown (ref. 7) that (at least in the low orders in which the calculations were made) the resultant scattering matrix elements coincided exactly with those obtained from the conventional theory after renormalization. No divergences are encountered in these calculations, as is not surprising, since one deals only with well-defined convolution products of distributions. Thus, the theory outlined here can replace the conventional quantum electrodynamics, putting in its place a divergence free formalism whose physical predictions are identical with those obtained before.

Theories of type $N > 1$ are a different matter. Here one encounters first a new free-field equation ($K^N a=0$). This does not cause any difficulties. The physical objects are the particles, as represented by the algebra of creation and annihilation operators in p-space and the associated Fock space. The fields $a(x)$ are certain linear homogeneous functions of these operators which are chosen more for the sake of the theory than for their physical meaning. The $a(x)$ are used as a basis with respect to which all operators of the theory are expanded. They also permit an easy way to express the spin-statistics connection. The wave packets (test functions on the mass shell) do of course have direct physical significance and occur in a careful evaluation of the S-matrix.

Now we do have a fundamental equation of the theory (e.g. in the form of the current equation), but pending a better method of solving it we resort to perturbation expansion. It now appears that if any order n of this expansion is to be computed, N can be suitably chosen (above a certain minimum) such that all convolution products exist. Up to order n all S-matrix elements can then be found without divergence difficulties. This means that one has a way of computing so-called nonrenormalizable theories in perturbation expansion. This method (originally in the form of replacing the P_{12} by the P_{12}^{N}) was first suggested by Chen (ref. 21). The difficulty with this solution (as one can show) is that one requires an increasing number of constants with increasing order of perturbation approximation in order to satisfy the boundary conditions. For this reason the perturbation solution does not seem to be physically meaningful for nonrenormalizable theories. Nevertheless, we believe that this method is of some value for two reasons: one is that the first few orders may nevertheless give reasonable approximations (e.g. for weak interactions), and the other is that mathematically it may indicate the way in which a solution of the current equation may have to be sought, viz. by going from simple polynomial type to transcendental type generalized functions, i.e. to non-tempered distributions.

Apart from these perturbation expansions it is of considerable interest to realize that we now have a quantum field theory in which there is no interpolating field at all (in the current-formulation) and in which the only field operators which occur besides the free fields and the S-operator are the currents. These correspond to the sources of the interpolating field in the field formalism, but they are not written as products of interpolating fields (such as $J_\mu(x) = \bar{\psi}\gamma_\mu\psi$). Thus we have a quantum field theory based entirely on free fields and sources. The latter are obtained from the S-operator by operator derivatives with respect to these free fields.

The following list of references does not claim to be complete, even for the period of the last few years, but it is hoped that all papers relevant to the present lectures are included.

References (in Chronological Order)

1. N. N. Bogoljubov, Izv. Akad. Nauk SSSR, Ser. Fiz. 19, 237 (1955).
2. H. Lehmann, R. Symanzik and W. Zimmermann, Nuovo Cimento 1, 205 and 2, 425 (1955); 6, 319 (1957).
3. V. Glaser, H. Lehmann and W. Zimmermann, Nuovo Cim. 6, 1122 (1957).
4. N. N. Bogoljubov and D. V. Shirkov, Introduction to the Theory of Quantized Fields, Interscience, New York 1959.
5. M. Muraskin and K. Nishijima, Phys. Rev. 122, 331 (1961).
6. V. Ya. Fainberg, Soviet Physics, JETP, 13, 1237 (1961).
7. R. E. Pugh, Ann. Phys. 23, 335 (1963).
8. F. Rohrlich, J. Math. Phys. 5, 324 (1964).
9. R. E. Pugh, Ann. Phys. 30, 422 (1964).
10. V. Ya. Fainberg, Z. Eksp. i Teor. Fiz. 47, 2285 (1964) - Soviet Physics JETP, 20, 1529 (1965).
11. F. Rohrlich and J. C. Stoddart, J. Math. Phys. 6, 495 (1965).
12. B. V. Medvedev, Z. Eksp. i Teor. Fiz. 48, 1479 (1965) - Sov. Phys. JETP, 21, 989 (1965).
13. R. E. Pugh, J. Math. Phys. 6, 740 (1965).

14. F. Rohrlich and F. Strocchi, Phys. Rev. 139, B476 (1965).
15. R. E. Pugh, J. Math. Phys. 7, 379 (1966).
16. F. Rohrlich and M. Wilner, J. Math. Phys. 7, 482 (1966).
17. T. W. Chen, F. Rohrlich and M. Wilner, J. Math. Phys. 7, 1365 (1966).
18. F. Rohrlich and J. G. Wray, J. Math. Phys. 7, 1697 (1966).
19. T. W. Chen, Nuovo Cim. 45, A533 (1966).
20. F. Rohrlich, "Asymptotic Quantum Field Theory", p. 295 in Perspectives in Modern Physics, ed. by R. E. Marshak, Interscience - Wiley, 1966.
21. T. W. Chen, Ann. Phys. (in press).
22. V. Gorgé and F. Rohrlich, J. Math. Phys. (in press).
23. J. G. Wray, Ph. D. Thesis, Syracuse University (1967)

In addition to these references the book by I. M. Gelfand and G. E. Shilov, Generalized Functions, Vol.I, Academic Press, New York, 1964, or a similar reference should be consulted for the mathematical background on generalized functions used here, and J. M. Jauch and F. Rohrlich , Theory of Photons and Electrons, should be consulted for notation and also for a presentation of a conventional quantum field theory.

QUANTUM FIELD THEORY IN DE SITTER SPACE†

By

W. THIRRING

Institute for Theoretical Physics
University of Vienna, Austria

1. Introduction

Usually elementary particle physics is based on the Poincaré group although the latter may be only the limiting form of the actual invariance group of the universe. One may have the feeling that cosmological changes of the space-time group will not change its microscopical consequences. On the other hand by contraction some important restrictions may be lost: going from the Poincaré group to the Galilei group the TCP - theorem or the connection between spin and statistics are lost. These restrictions are important even when apparently no relativistic velocities are involved. Similarly the difference between the Poincaré group and the de Sitter group could show up in laboratory experiments. Consequently we shall investigate how quantum field theory works for systems invariant under the de Sitter group. We do not mean to imply that we consider as established that we

† Lecture given at the VI. Internationale Universitätswochen f.Kernphysik,Schladming,26 February-11 March 1967.

live in a de Sitter space but rather consider two different models of space-time, both invariant under the de Sitter group, as interesting possible representations of our cosmos. Furthermore we shall not discuss all relevant questions but concentrate on one main difference between Poincaré and de Sitter group, namely the definiteness of the energy. This already shows up in the reduced model (one space, one time dimension) and for a scalar field. Thus for simplicity I shall restrict myself to this case and refer to a paper by O. Nachtmann [1] for a discussion of the full model and of spin 1/2 fields. We shall use the oldfashioned (or time-honored) Lagrangian approach [2] and find that it works for the de Sittergroup just as well as for the Lorentz group. Only in the model where space is finite but expanding something unusual happens. An interaction introduces a continuous creation of particles. But for a reasonably large universe the rate is so small that it is irrelevant and no microscopically significant difference between Poincaré and de Sitter space is visible.

2. Representations

The invariance group of the reduced model leaves the surfaces

$$x_0^2 - x_1^2 - x_2^2 = \pm R^2 \tag{1}$$

invariant and is thus O(2,1). (It is locally isomorphic to SL(2,R) and SU(1,1)). It has three infinitesimal generators K_i whose commutation relations differ by a sign from the familiar ones of angular momentum:

$$[K_1, K_2] = -iK_3, [K_2, K_3] = iK_1, [K_3, K_1] = iK_2 \tag{2}$$

There is one invariant

$$I = K_3^2 - K_1^2 - K_2^2 \tag{3}$$

We shall first consider unitary representations where the K_i are self-adjoint operators. In a representation where K_3 is diagonal*

$$(m|K_3|m') = m\delta_{mm'} \tag{4}$$

we have for

$$K_\pm = K_1 \pm K_2 \,, \quad K_3 K_\pm = K_\pm(K_3 \pm 1) \tag{5}$$

the relation

$$K_\pm K_\mp = - I + m(m \mp 1) \geq 0 \tag{6}$$

Thus either $I \leq 0$ or the sequence

$$K_\pm|m) = \sqrt{-I+m(m\pm1)}\ |m\pm1) \tag{7}$$

terminates for an upper m_u or a lower m_ℓ :

$$K_-|m_\ell) = 0 : I = m_\ell(m_\ell - 1)$$

$$K_+|m_u) = 0 : I = m_u(m_u + 1) \tag{8}$$

In summary for unitary representations K_3 and I have the following eigenvalues (fig. 1).

---

* since K_3 is the rotation in the x_1-x_2-plane m,m' are integers. We shall not consider representations of the covering group which gives half integer m.

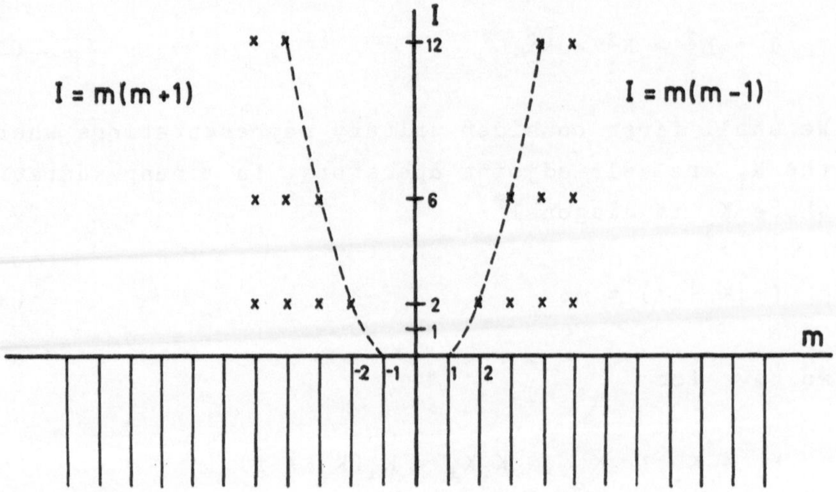

Fig. 1

There are two different contractions*:

1. $O(2,1) \to O(2) \overset{s}{\cdot} P_2$

This corresponds to the limit $R \to \infty$ of $P_{1,2} = K_{1,2}/R$ which gives the 2-dimensional Euclidian group:

$$(K_3, P_\ell) = i\varepsilon_{\ell K} P_K \ , \quad (P_1, P_2) = 0 \qquad (9)$$

Geometrically it corresponds to projection of the surface (1) with the upper sign (fig. 2), onto a tangential plane. Since the latter carries Euclidian metric it cannot serve as a space-time model.

2. $O(2,1) \to O(1,1) \overset{s}{\cdot} P_2$

Here we have:

$$(P_1, P_3) = 0, \quad (K_2, P_1) = iP_3$$

$$(K_2, P_3) = iP_1$$

* P_2 denotes a 2-dimensional translation group and $\overset{s}{\cdot}$ is a semi-direct product.

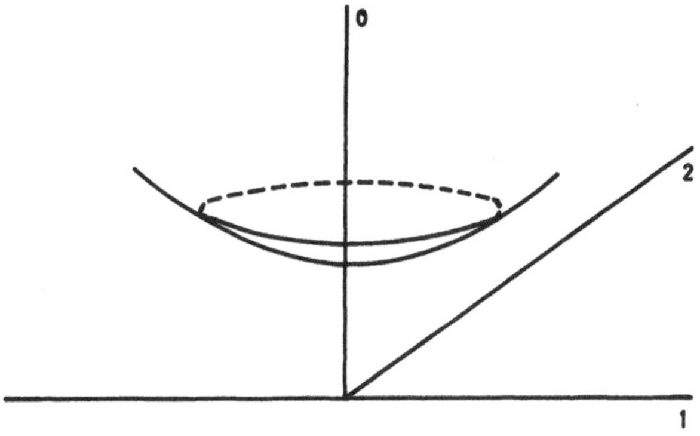

Fig. 2

$$P_{1,3} = K_{1,3}/R \quad , \; I \to R^2(P_3^2 - P_1^2) = R^2\mu^2 \tag{10}$$

In the limit $R \to \infty$ P_3 obtains a continuous spectrum with "eigenvalues" $p_3 = m/R$:

$$(m|P_3|m') = p_3 \, \delta_{mm'} \tag{11}$$

From (7) we see that for the other operators the representations become

$$P_1 |m) = p_1 \frac{1}{2}(|m + 1) + |m - 1)) \; , \; p_1 = \sqrt{p_3^2 - \mu^2}$$

$$K_2 |m) = p_1 \frac{R}{2i}(|m + 1) - |m - 1)) \tag{12}$$

Thus P_1 has eigenvalues $\pm p_1$ whereas K_2 is represented by

$$p_1 \frac{1}{i} \frac{d}{dp_3} = \frac{1}{i}(p_1 \frac{\partial}{\partial p_3} + p_3 \frac{\partial}{\partial p_1})$$

since

$$\frac{\partial p_1}{\partial p_3} = \frac{p_3}{p_1}$$

We have to distinguish two cases:

a) I, $\mu^2 > 0$, $\rightarrow O(3,2)$

Here the eigenvalues are restricted by

$$\mu < p_3 < \infty \quad , \quad -\infty < p_1 < \infty \tag{13}$$

like a particle with mass μ if p_3 is the energy and p_1 the momentum. This corresponds to a projection of the surface (1) with the lower sign onto a tangential plane where time t and space x are the following directions:

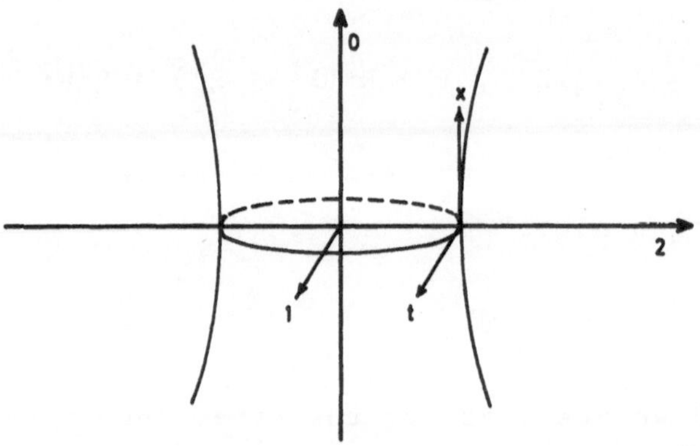

Fig. 3

Thus it represents a universe where space is infinite and time periodic. Here we have the analogue of representations where the energy is always positive.

b) I, $\mu^2 < 0 \rightarrow O(4,1)$

Here the eigenvalues are

$$-\infty < p_3 < \infty \ , \ |\mu| < p_1 < \infty \ \text{ and } \ -\infty < p_1 < -|\mu| \ .$$

Here we have the usual energy-momentum relation with real mass if p_1 is the energy and p_3 the momentum. Hence in Fig. 3, x and t exchange their role. Thus this is a different universe which is infinite in time and periodic in space. Here the energy is not always positive which is connected with the fact that K_1 shifts the surface (Fig. 3) up for $x_2 > 0$ and down for $x_2 < 0$. Thus particles on the right hand side will correspond to positive K_1 and on the left hand side to $K_1 < 0$.

3. Conservation Laws

We shall investigate a free scalar field which is described in general coordinates by the Lagrangian density

$$L = \frac{1}{2} \sqrt{-g} \, (\Phi_{,i} \Phi_{,K} \, g^{iK} - \mu^2 \Phi^2) \tag{14}$$

The energy-momentum-tensor density is

$$T_{iK} = \sqrt{-g} \, \Phi_{,i} \Phi_{,K} - g_{iK} \, L \tag{15}$$

and satisfies

$$T^{K}_{i,K} = 0 \tag{16}$$

A conservation law with an ordinary rather than a covariant derivative is obtained if one finds coordinate transformations $x \rightarrow x + \delta x$ which leave g^{iK} form-invariant:

$$\delta^* g^{iK} = \delta x^{i}_{,m} \, g^{mK} + \delta x^{K}_{,m} \, g^{im} - g^{iK}_{,m} \, \delta x^{m} = 0 \tag{17}$$

For these one has

$$(\delta x^{K} \, T^{m}_{K})_{,m} = 0 \tag{18}$$

which then can be written in the integrated form

$$\int \delta x^K \, T_K^m \, d\sigma_m = \text{const.} \tag{19}$$

Furthermore for variations $\bar{\delta}\Phi$ of Φ which leave L invariant (e.g. phase changes if Φ is a complex field) one gets a conserved current density:

$$I^K = \bar{\delta}\Phi \frac{\delta L}{\delta \Phi_{,K}} \; , \quad I^K_{\;,K} = 0 \quad , \quad \int d\sigma_K \, I^K = \text{const.} \tag{20}$$

These general formulae will now be evaluated for the reduced model. Introducing in (1) the coordinates

$$x_o = R\,\mathrm{Sh}\chi \quad , \; x_1 = R\,\mathrm{Ch}\chi\,\cos\phi \; , \quad x_2 = R\,\mathrm{Ch}\chi\,\sin\phi \tag{21}$$

we find

$$g_{iK} = R^2 \begin{bmatrix} 1 & 0 \\ 0 & -\mathrm{Ch}^2\chi \end{bmatrix} \begin{matrix} \chi \\ \phi \end{matrix} \qquad \begin{matrix} \chi & \phi \end{matrix} \tag{22}$$

Thus

$$L = \frac{1}{2}\,\mathrm{Ch}\chi\,(\Phi^2_{,\chi} - \mathrm{Ch}^{-2}\chi\,\Phi_{,\phi} + R^2\mu^2\Phi^2)$$

$$T_i^K = \begin{bmatrix} \frac{1}{2}(\mathrm{Ch}\chi\,\Phi^2_{,\chi} + \mathrm{Ch}^{-1}\chi\,\Phi^2_{,\phi} - R^2\mu^2\mathrm{Ch}\chi\,\Phi^2) \; ; & -\Phi_{,\chi}\Phi_{,\phi}\mathrm{Ch}^{-1}\chi \\ \Phi_{,\chi}\Phi_{,\phi}\mathrm{Ch}\chi \; ; & -\frac{1}{2}(\mathrm{Ch}\chi\,\Phi^2_{,\chi} + \mathrm{Ch}^{-1}\chi\,\Phi^2_{,\phi} + R^2\mu^2\mathrm{Ch}\chi\,\Phi^2) \end{bmatrix} \tag{23}$$

Thus for ϕ timelike and $\mu^2 > 0$ or χ timelike and $\mu^2 < 0$ the energy density is positive definite. However, since the conservation law (16) is

$$T^\chi_{\chi,\chi} + T^\phi_{\chi,\phi} = -\,\mathrm{Ch}\chi\,\mathrm{Sh}\chi\,T^{\phi\phi}$$

$$T^{\chi}_{\phi,\chi} + T^{\phi}_{\phi,\phi} = 0 \tag{24}$$

one obtains only in the former case a positive definite integrated quantity which is conserved. Furthermore the three generators of the de Sitter group bring three Killing-vectors δx which give via (19) conserved integrated quantities. They are

$$K_3 : \delta x = \varepsilon \binom{0}{1} \begin{matrix}\chi \\ \phi\end{matrix},$$

$$\delta x^K T^o_K = -\frac{R^{-2}}{2}(\mathrm{Ch}\chi \Phi^2_{,\chi} + \mathrm{Ch}^{-1}\chi \Phi^2_{,\phi} + R^2\mu^2 \mathrm{Ch}\chi \Phi^2)$$

$$K_2 : \delta x = \varepsilon \binom{\cos\phi}{-\sin\phi \ \mathrm{Tg}\chi},$$

$$\delta x^K T^o_K = \frac{R^{-2}}{2} \cos\phi(\mathrm{Ch}\chi \Phi^2_{,\chi} + \mathrm{Ch}^{-1}\chi \Phi^2_{,\phi} - R^2\mu^2 \mathrm{Ch}\chi \Phi^2) -$$

$$- \sin\phi \ \mathrm{Sh}\chi \ \Phi_{,\chi} \ \Phi_{,\phi} \ R^{-2}$$

$$K_1 : \delta x = \varepsilon \binom{\sin\phi}{\cos\phi \ \mathrm{Tg}\chi},$$

$$\delta x^K T^o_K = \frac{R^{-2}}{2} \sin\phi(\mathrm{Ch}\chi \Phi^2_{,\chi} + \mathrm{Ch}^{-1}\chi \Phi^2_{,\phi} - R^2\mu^2 \mathrm{Ch}\,\chi \Phi^2) +$$

$$+ \cos\phi \ \mathrm{Sh}\chi \ \Phi_{,\chi} \ \Phi_{,\phi} \ R^{-2} \tag{25}$$

Thus, for ϕ = timelike K_3 is the time displacement and again for $\mu^2 > 0$ we have a definite energy. For χ = timelike K_1 is the local time displacement which gives a conserved integrated quantity but even for $\mu^2 < 0$ it is not positive definite. The complex solutions of the wave equation

$$\mathrm{Ch}\chi \ \Phi_{,\chi\chi} + \mathrm{Sh}\chi \ \Phi_{,\chi} - \mathrm{Ch}^{-1}\chi \ \Phi_{,\phi\phi} - R^2\mu^2 \mathrm{Ch}\chi \ \Phi = 0 \tag{26}$$

which follows from L give via (20) a conserved vector density. Thus we have the invariants

a) $N = i\int d\chi \; Ch^{-1}\chi \; (\Phi^{*} \Phi_{,\phi} - \Phi^{*}_{,\phi} \Phi)$

b) $N = i\int d\phi \; Ch\chi \; (\Phi^{*} \Phi_{,\chi} - \Phi^{*}_{,\chi} \Phi)$ (27)

in our two cases.

4. Quantization

In this section we shall study the consequences of the canonical quantization prescription for our system. To ease the algebra we shall from now on use units such that $R = 1$. Again we have to distinguish the two cases

a) ϕ = time, χ = space.

The canonical conjugate field is

$$\Pi(\phi,\chi) = Ch^{-1}\chi \; \Phi_{,\phi}(\phi,\chi) \tag{28}$$

and thus the canonical commutation rules are

$$\left[\Phi(\phi,\chi), \Phi_{,\phi}(\phi,\chi')\right] = i \; Ch\chi \; \delta(\chi - \chi')$$

$$\left[\Phi(\phi, \chi), \Phi(\phi,\chi')\right] = \left[\Phi_{,\phi}(\phi,\chi), \Phi_{,\phi}(\phi,\chi')\right] = 0 \tag{29}$$

They imply that three generators of the de Sitter group are actually given by integrating the densities (25) over space:

$$K = \int_{-\infty}^{\infty} d\chi \; \delta x^{K} \; T^{o}_{K}$$

$$(K_3,\Phi) = i \; \Phi_{,\phi}$$

$$(K_{\pm},\Phi) = e^{\mp i\phi}(\pm\Phi_{,\chi} - i \; Tg\chi \; \Phi_{,\phi}) \tag{30}$$

The differential operators on the right hand sides

correspond in fact to the infinitesimal transformations K_i.

b) χ = time, ϕ = space.

Here the conjugate field is

$$\Pi(\chi,\phi) = \mathrm{Ch}\chi \; \Phi_{,\chi}(\chi,\phi) \tag{31}$$

implying the commutation relations

$$\left[\Phi(\chi,\phi) \; , \; \Phi_{,\chi}(\chi,\phi')\right] = \mathrm{Ch}^{-1}\chi \; \delta(\phi - \phi')$$

$$\left[\Phi(\chi,\phi) \; , \; \Phi(\chi,\phi')\right] = \left[\Phi_{,\chi}(\chi,\phi), \; \Phi_{,\chi}(\chi,\phi')\right] = 0 \tag{32}$$

Again the generators of the de Sitter group with the commutation relations (30) are given by integrating (25) over space.

$$K = \int_0^{2\pi} d\phi \; \delta x^j \; T_j^o \tag{33}$$

One might wonder whether this quantization prescription is independent of our choice of coordinates. Whereas in flat space there are preferred reference frames it is here a matter of taste which coordinates one chooses [3]. It would be embarrassing if the canonical prescription did not lead to equivalent results in different systems. We shall show that this is not the case by exhibiting the generator E of an infinitesimal coordinate transformation $x \to \bar{x} = x+\delta x$. The latter is in fact given by the usual expression:

$$E = \int dx_1 \; \delta x^K \; T_K^o \tag{34}$$

where the δx^K are arbitrary functions of the coordinates. E transforms Φ like a scalar and Π like a zero component of a contravariant vector density. This is what happens if we introduce $\bar{x}$ in L since

$\Pi = +\sqrt{-g}\,\Phi_{,i}\,g^{io}$. Thus the canonically conjugated quantities in the new frame are obtained from the old ones by a unitary transformation and thus the canonical commutation relations in the two frames are equivalent. We shall demonstrate this transformation property of E only for the case b) where it becomes ($\Pi = \mathrm{Ch}\chi\Phi_{,\chi}$)

$$E = \frac{1}{2}\int_{o}^{2\pi} d\phi\left((\Phi^2_{,\chi}+\mathrm{Ch}^{-2}\chi\Phi^2_{,\phi}-\mu^2\Phi^2)\delta\chi+\Phi_{,\chi}\Phi_{,\phi}\delta\phi\right)\mathrm{Ch}\chi \quad (35)$$

One finds

$$\delta\Phi = i(\Phi,E) = -\Phi_{,\chi}\delta\chi-\Phi_{,\phi}\delta\phi = -\delta x^i\Phi_{,i}$$

$$\begin{aligned}\delta\Pi = i(\Pi,E) &= -\frac{\partial}{\partial\phi}(\mathrm{Ch}^{-1}\chi\Phi_{,\phi}\delta\chi) - \mu^2\Phi\mathrm{Ch}\chi\delta\chi - \\ &\quad - \frac{\partial}{\partial\phi}(\mathrm{Ch}\chi\Phi_{,\chi}\delta\phi) = \\ &= -\delta\chi(\mathrm{Ch}^{-1}\chi\Phi_{,\phi\phi}+\mu^2\mathrm{Ch}\chi\Phi) - \mathrm{Ch}^{-1}\chi\Phi_{,\phi}\delta\chi_{,\phi} - \\ &\quad - \mathrm{Ch}\chi\Phi_{,\chi}\delta\phi_{,\phi} - \mathrm{Ch}\chi\Phi_{,\chi\phi}\delta\phi = \\ &= -\delta x^i\Pi_{,i} - \delta x^i_{,i}\Pi + \sqrt{-g}\,\delta x^o_{,j}\,g^{jK}\Phi_{,K} \quad (36)\end{aligned}$$

in agreement with the above mentioned transformation properties.

5. Creation and Destruction Operators

Finally we shall come to an explicit representation of the commutation relations in terms of creation and destruction operators. For the Poincaré group this is effected by a decomposition of the field in positive and negative frequency parts which is invariant under orthochronous transformations. In de Sitter space there are no preferred time coordinates and thus the question arises how one can find a decomposition which is independent of the coordinate system one happens to use. We shall do this with the aid of those complex solutions of the wave equation (26) which correspond to the positive or negative eigenvalues of the invariant (27) [4] in a suitable Hilbert space. Like the angular part of the Laplace equation, (26) can be solved in terms of Legendre functions except that now the argument is i Shχ. The result is that the solutions with the desired properties expressed in terms of the hypergeometric function F are the following:

a) ϕ = time, χ = space

Here solutions which stay bounded everywhere exist only if the mass assumes certain quantized values $\mu^2 = n(n+1)$, n = integer, corresponding to the values of the invariant for this case.

$$q_n^m(\chi) =$$

$$= (-)^m \frac{\sqrt{\Gamma(m+n+1)}}{\sqrt{\Gamma(m-n)}} \frac{\sqrt{m}}{\sqrt{n+1}} \frac{2^{n+1}\ \Gamma(n+2)}{\sqrt{\pi\Gamma(2n+2)\Gamma(2n+3)}} \frac{\mathrm{Ch}^m\chi}{\mathrm{Sh}^{m+n+1}\chi} \cdot$$

$$\cdot\ F\left(\frac{m+n}{2} + 1,\ \frac{m+n}{2} + \frac{1}{2},\ n + \frac{3}{2},\ -\frac{1}{\mathrm{Sh}^2\chi}\right) \qquad (37)$$

$$\Phi(\chi,\phi) = \sum_{m=n+1}^{\infty} \left(e^{-im\phi} q_n^m(\chi) \frac{a_m}{\sqrt{2m}} + e^{im\phi} q_n^m(\chi) \frac{a_m^\dagger}{\sqrt{2m}}\right) \qquad (38)$$

If the creation and destruction operators $a^\dagger$ and a satisfy the standard commutation relations

$$[a_m, a_{m'}^\dagger] = \delta_{mm'} \qquad (39)$$

one can show with formulas of classical analysis that (29) is satisfied. Furthermore by contracting one sees that (38) goes over into the usual expansion for a free scalar field, q_n^m reducing to sin x or cos x.

b) χ = time, ϕ = space.

Here the functions have to be normalized somewhat differently. μ^2 may now assume the negative values $-\frac{1}{4} - \lambda^2$, λ real.

$$\tilde{q}^m_{i\lambda-1/2} = \frac{(-)^m}{\sqrt{2\lambda}} (2\lambda)^{-i\lambda} \frac{\sqrt{\Gamma(m+i\lambda+1/2)}}{\sqrt{\Gamma(m-i\lambda-1/2)}} \frac{\mathrm{Ch}^m \chi}{\mathrm{Sh}^{m+1/2+i\lambda}} \cdot$$

$$\cdot\ F\left(\frac{m+i\lambda}{2} + \frac{3}{4}, \frac{m+i\lambda}{2} + \frac{1}{4}, 1+i\lambda\ , -\frac{1}{\mathrm{Sh}^2\chi}\right) \quad (40)$$

$$\Phi(\chi,\phi) = \sum_{m=-\infty}^{\infty} \{e^{-im\phi}\tilde{q}^m_{i\lambda-1/2}(\chi)a_m + e^{im\phi}\tilde{q}^m_{-i\lambda-1/2}(\chi)a^\dagger_m\}$$

Again if a and $a^\dagger$ obey (39), the canonical commutation rules are satisfied. Using classical properties of F one can work out the generators (30) of the de Sitter group. In both cases they can be written

$$K_3 = \sum_m a^\dagger_m a_m m$$

$$K_\pm = \sum_m a^\dagger_m a_{m\mp1} \sqrt{m(m\mp1)-\mu^2} \quad (41)$$

In case a) ($\mu^2>0$) the sum over m goes from n+1 to ∞ so that K_3 is positive. In case b) ($\mu^2<0$) m goes from $-\infty$ to ∞ in agreement with the general analysis of the representations of O(2,1). The invariant number of particles operator is obtained by inserting into (27) for Φ only the part with a_m and for Φ^* its Hermitian conjugate. In both cases it becomes

$$N = \sum_m a^\dagger_m a_m \quad (42)$$

6. Reflexions

So far we considered only the connected part of the group. It remains to discuss whether the de Sitter group offers some new aspects regarding discrete operations. We shall see that no essential effects turn up as is intuitively expected from the reflexion symmetry of the hyperboloid in fig. 3. However one has to keep in mind that parity and time reversal are local concepts, i.e. they refer to an arbitrarily chosen coordinate origin. In the two models they have the following representations:

a) ϕ = time, χ = space.

Here the energy is given by K_3 and the momentum by K_1 or K_2, depending on where one is in the universe.

1. Parity:

It corresponds to

$$x_3 \to -x_3 \ , \ x_{1,2} \to x_{1,2} \ , \text{ or } \chi \to -\chi, \ \phi \to \phi \tag{43}$$

Using the properties

$$q_n^m(-\chi) = (-)^{m-n-1} \ q_n^m(\chi) \tag{44}$$

we see that the unitary operator P_a with

$$P_a \ a_m \ P_a^{-1} = (-)^{m-n-1} \ a_m \tag{45}$$

has the desired properties

$$P_a \ \Phi(\phi,\chi) \ P_a^{-1} = \Phi(\phi,-\chi) \ . \tag{46}$$

Furthermore we see that

$$P_a K_3 P_a^{-1} = K_3 \ , \quad P_a K_{1,2} P_a^{-1} = -K_{1,2} \tag{47}$$

as is expected for a parity operation. It is interesting to note that P_a combined with a rotation around the 3-axis by π,

$$\theta_a = P_a e^{i\pi K_3} \tag{48}$$

is a unitary reflexion of all coordinates, i.e. $x_i \to -x_i$:

$$\theta_a \Phi(\phi,\chi)\theta_a^{-1} = \Phi(\phi+\pi,-\chi) \ , \quad \theta_a K_i \theta_a^{-1} = K_i \tag{49}$$

2. Time reversal

Since the q_n^m are real functions the transformation

$$x_2 \to -x_2 \ , \ x_{1,3} \to x_{1,3}, \text{ or } \chi \to \chi, \ \phi \to -\phi \tag{50}$$

is effected by the antiunitary operation K_a with

$$K_a \, a_m \, K_a^{-1} = a_m \ :$$

$$K_a \Phi(\phi,\chi) \, K_a^{-1} = \Phi(-\phi,\chi) \quad ,$$

$$K_a K_{1,3} K_a^{-1} = K_{1,3} \ , \ K_a K_2 K_a^{-1} = -K_2 \tag{51}$$

Hence there is no difficulty in this case in defining improper operations.

b) χ = time, ϕ = space.

Here K_3 is the momentum and K_2, for instance, can be considered as energy.

Hence we have

1. Parity:

$$x_2 \to -x_2 \ , \ x_{1,3} \to x_{1,3} \ , \quad \phi \to -\phi \ , \ \chi \to \chi \ .$$

Our $\tilde{q}_n^m$ have the property

$$\tilde{q}^{-m} = (-)^m \tilde{q}^m \tag{52}$$

and thus

$$P_b \, a_m \, P_b^{-1} = (-)^m \, a_{-m} \tag{53}$$

generates this parity operation:

$$P_b \, \Phi(\chi,\phi) \, P_b^{-1} = \Phi(\chi,-\phi) \tag{54}$$

$$P_b K_2 P_b^{-1} = K_2 , \qquad P_b K_{1,3} P_b^{-1} = -K_{1,3}$$

2. Time reversal

$$x_3 \to -x_3 , \quad x_{1,2} \to x_{1,2} , \text{ or } \chi \to -\chi , \ \phi \to \phi$$

The $\tilde{q}_n^m$ as defined in (40) are not well chosen for expressing this operation. Instead one has to use q's with

$$\tilde{q}_n^m(\chi)^* = \tilde{q}_n^{-m}(-\chi) \quad . \tag{55}$$

In terms of these the anitunitary K_b with

$$K_b a_m K_b^{-1} = a_{-m} \tag{56}$$

yields

$$K_b \Phi(\chi,\phi) K_b^{-1} = \Phi(-\chi,\phi)$$

$$K_b K_3 K_b^{-1} = -K_3 , \qquad K_b K_{1,2} K_b^{-1} = K_{1,2} \, . \tag{57}$$

Summarizing we may say that there is no essential difference between flat and curved space regarding reflexions. The unitary reflexion (48) does not correspond to the usual (antiunitary) TCP - operation since in no local frame it has the property $x \to -x$, $t \to -t$.

7. Discussion

Our development so far has shown that the canonical formalism works for the de Sitter group just as well as for the Poincaré group. Only in case b) (χ = time) there is the difference that K_1 (the local time displacement operator) is no longer positive definite. One sees easily that a state $\sum_m a_m^\dagger|0\rangle$ is approximately an eigenstate with positive eigenvalue and $\sum_m (-)^m a_m^\dagger|0\rangle$ one with negative eigenvalue. The former is concentrated around $\phi \sim -\pi/2$, the latter around $\phi \sim \pi/2$. This is in agreement with our earlier observation that K_1 shifts the time up at one side and down at the other. The question is whether in this representation the three constants K_i guarantee a kinematically stable situation. Of course, for free fields nothing happens but with a $G\Phi^3$-interaction one finds that spontaneous creation is possible and happens indeed. However, the rate is very low [4], (in 3+1 dimensions) per space-time volume $\sim G^2/R^2\ e^{-R|\mu|}$. As to be expected it has the characteristic factor $e^{-R|\mu|}$ corresponding to the tunneling of a virtual particle from one side of the universe to the other. Even for $\mu = 0$ the rate is insignificant for practical purposes. Thus there does not seem to be any argument from elementary particle physics against either of the world models.

References

1. O.Nachtmann, to be published in Communications of Mathematical Physics.

2. For other attempts see: C. Fronsdal, Rev. Mod. Phys. 37, 221 (1965); P. Roman, J. J. Aghassi, Nuovo Cim. 152, 193 (1966).
3. Compare E. Schrödinger, Expanding Universes, Oxford 1956.
4 O Nachtmann, to be published.

MACH'S PRINCIPLE AND ELEMENTARY PARTICLES†

By

O. BERGMANN

George Washington University
Washington D. C., USA

1. Introduction, Mach's Principle in Classical Physics

There are at least three different formulations of this principle which is named after Ernst Mach. The most specific is in terms of the Foucault pendulum which, according to the principle [1], moves in a constant plane with respect to the fixed stars, rather than to the absolute space, as Newton had thought. The second formulation, obviously closely related, but more general than the first, is the statement by Mach, that all motion is motion relative to other bodies [2]. Finally, one may consider the statement that inertia depends on the other bodies in the universe, as Mach's principle, although it was formulated in this general form by Einstein [3]. The three formulations emphasize different aspects of Mach's principle, and therefore require different criticism. It is clear that rotational motions considered in the first for-

---

† Lecture given at the VI. Internationalen Universitätswochen f.Kernphysik,Schladming,26 February-11 March 1967.

mulation will appear differently when the theory of relativity is taken into account; one need only recall the problem of the rotating disk. According to Newtonian ideas, two Cartesian coordinate frames in rotation relative to each other are "metrically" equivalent, since simultaneity has the same meaning for both observers, and distances of simultaneous events will thus be expressed in the same way. But the two observers are not dynamically equivalent, since in one or in both frames inertial forces will appear. A priori, one could think of two specific coordinate frames; one in which the Foucault pendulum remains in a constant plane, and another one in which the fixed stars remain at rest and it seems quite remarkable that these frames are in fact identical.

In the special theory of relativity there is no equivalence in any sense between rotating coordinate frames, and it seems therefore less remarkable that the only distinguished frame has both features in common. The general theory of relativity supplies a method for calculation of the metric tensor for a given energy-momentum distribution, and some specified boundary conditions. This is indeed a great step towards a solution of Mach's principle, because the knowledge of the metric tensor permits the calculations of the inertial forces at every point of space-time.

However, if the imposed boundary conditions required the flatness of space-time at infinity, one would have returned to the absolute space-time concept, admittedly objectionable now more on philosophical than on physical grounds [4]. In any case Einstein proposed a spatially closed universe to abolish the absolute space completely.

This aspect of Mach's principle has then found an at least qualitative solution. However, merely from the usefulness of the Schwarzschild solution, which represents the gravitational field of an isolated mass point, we may conclude that the determination of the distinguished frame is largely insensitive to the actual distribution of all other masses in the universe. And one should add that there is evidence of a motion of our galaxy. Obviously the final answer to this part of Mach's principle will be found when the many body problem in the general theory of relativity can be solved, and more accurate data on the motion of stars and the local inertial frame are available. We will return to this point shortly.

The second formulation appears more philosophical than physical. There has never been a theory in which this demand for the exclusive use of relative coordinates is satisfied [5], because the occurrence of coordinates which can be transformed is already in violation of this principle, at least in spirit.

In fact the general theory of relativity with its use of general covariance seems to have removed us even further from the goal, because a local coordinate transformation will make a body appear to move differently, and though one may claim that the difference in the motion of this body in the two frames is unobservable,such coordinate transformations are useful in cases where the reference bodies which serve as coordinate frames do not significantly affect the motion of the body to be observed. In any case, W. Thirring [6] pointed out that the formulation of the problem rests on a classical interpretation of space. In the modern language, one should not hesitate to say that a particle moves relative to all points of space since these points are potentially occupied (with some probability) by other particles.

While the space even in Einstein's time was an empty metrical manifold, and it was a constant worry to imagine a physical space with one or with no particles, modern theory would reserve the concept of a physical space for a manifold with certain fields defined on it, capable of describing particles according to the rules of quantum theory. But again, the answer is hardly satisfactory, if the field theory now defines an absolute space because, however deceptive macroscopic observation may be, the Foucault pendulum requires an explanation also.

We cannot give here a complete discussion of the third and most general formulation of Mach's principle. It is a special expression of the general demand to "complete" any physical theory, but it is too vague to permit a thorough analysis. Einstein [7] expected as a consequence of this dependence of the inertia of a body on the remaining bodies in the universe, that the inertial mass changes whenever the body is approached by other masses. Unfortunately, calculations to show that this is indeed the case appear unconvincing since the effect does not occur in a geodesic frame of references, and the rest mass is in no way affected, as one should have expected. There are attempts to demonstrate the effects of the anisotropy of the universe due to the inhomogeneity of the distribution of stars. One would expect different values of the curvature tensor, say in the direction of the galaxy, and perpendicular to this direction. The equation of the geodesic deviation contains the curvature tensor, and would allow us in principle to detect such anisotropy of space, but the effect is far too small to permit a test of Mach's principle.

It is often felt that the stars do not have a sufficient effect on our local phenomena but this may be misleading, since an influence with a long range could appear uniform over a region comparable with the extension of the planetary system, and thus be hard to observe locally. In any case, we may introduce at any point of space-time a geodesic coordinate frame, and it seems illogical to make the stars responsible for the appearance of inertial forces, if we do not use the geodesic frame; it is more appropriate to say that the inertial forces are due to the choice of the coordinate frame, since nothing else in the universe has been changed. In other words, although the local inertial frame is fixed by the distribution of the stars, if there were none,it would be "determined" by something else, viz. by the nature of the Riemannian manifold, which we have assumed from the outset. The general theory of relativity allows us only to calculate the local deviations in geometrical quantities from a basic Riemannian manifold which we have to assume. Einstein concluded therefore that "we must expect the whole inertia, that is, the whole g-field, to be determined by the matter of the universe..." [8].

A full appreciation of Einstein's statement is essential for understanding Mach's principle. Previously proposed models of Mach's principle start usually with a realization of long-range effects of masses, without explaining the metric tensor itself. One may describe long-range effects by a field as due to an accelerated source at large distances. This was suggested by Sciama [9]. The r^{-1} field will appear only when the sources are accelerated, and this is in accordance with the fact that inertial forces appear whenever a body is accelerated with respect to the fixed stars. We will not discuss this theory, in any detail, and will merely mention also a second model [10]. The coupling of a scalar field with point particles can be described most simply by the Lagrangian

$$L = V\sqrt{\dot{X}^n \dot{X}_n} \tag{1}$$

which suggests that the mass itself may be proportional to a scalar field. The observed discreteness of the masses must then result from quantum mechanics. The equation of motion resulting from (1) is

$$\ddot{X}^q - (\log V)_{,p}(g^{pq} - \dot{X}^q\dot{X}^p) = 0 \qquad (2)$$

Since the gravitational interaction is fully accounted for by the Christoffel symbols, we would like to argue that the last term is under ordinary circumstances negligible.

We may assume this to be true for the contributions from the distant stars, but the local masses require special consideration. Let us assume that V is a sum over all masses, more specifically,

$$V = \Sigma b/r_n \simeq V_o + b/r \qquad (3)$$

where V_o is a large constant representing the distant masses and b is a universal constant. In the non-relativistic approximation, we get, for the motion of a test particle, assuming the usual Schwarzschild solution due to the neighbouring mass,

$$\ddot{\underline{X}} = -(GM - b/V_o)\underline{X}/\underline{X}^3 \qquad (4)$$

Apart from the redefinition of the mass, this is again Newton's equation of motion, but we note that the perihelion motion of a test particle in the field of the heavy mass will differ from the Einstein prediction. According to the previously stated interpretation of V, we should now identify the constant of integration M from Einstein's equation with the sum of all contributions in the V-field, i.e.

$$M = V_o \qquad (5)$$

and the very well established identity of inertial and gravitational mass would then not hold in this model theory. In any case, the relation (5) can be meaningful only for elementary particles or in a theory with a whole set of possible values of V rather than a single value, that is in a quantized version of the theory.

A general formulation of the relation (5) is not easy to find. A single particle in the universe could not exist because its gravitational mass would have to be zero since there are no sources for the V-field, unless we allow for wave-solutions for V. It is here assumed that the mass of a particle is equal to that part of V produced by all other particles and there is then no self-energy problem. However, a coupling between the metric field and the V-field would occur in any reasonable execution of this model theory. Consequently, the Schwarzschild metric would probably be only an approximation. We will not discuss this naive model any further, but mention finally a serious weakness of the classical interpretation of Mach's principle. A model by Ozvath and Schücking [11] was called by the authors an "anti-Mach"-solution of the Einstein equations, a view which was criticized by Hönl and Dehnen [12]. The interpretation of this solution given by the latter authors is an Einstein universe with a superimposed gravitational wave. This interpretation is sufficient to show why the classical form of Mach's principle is too narrow. The classical theory distinguished unduly between mass and energy. Not only should e.g. radio stars contribute to the effects we seek to describe, but other energies like field energies including the energy of a gravitational wave, should also contribute. It seems therefore very likely that a complete solution of Mach's principle is beyond our present knowledge of the universe. We can look only for promising theoretical models [13]. The discussion in the last paragraph can be illustrated as follows. The only quantitative formulation of Mach's principle, known for a long time [14] is the relation

$$G\, m_u / r_u\, c^2 = 1 \qquad (6)$$

where m_u and r_u are the mass and the radius of the universe respectively. In the limit $m_u = 0$, the radius r_u will vanish, and this is a plausible consequence of Mach's principle. However, from a modern point of view, fields which describe mass-less particles are just as amenable to a mathematical description as other fields, and one should therefore replace m_u by the total energy of the universe, although this is by no means a logical necessity. If on the other hand we take the fictitious limit $G = 0$, the radius should vanish again, and this is somewhat surprising since the gravitational interaction is so extremely weak for elementary particles, and one might have thought that the strong interactions should be dominant here too. It would be highly speculative, and useless, to answer this question in one way or the other. The relationship of the masses of the elementary particles and their interaction parameters and cosmic quantities, i.e. the final touch to a theory of elementary particles, is certainly beyond our present knowledge and understanding and, what is more to the point under consideration, it is not part of Mach's principle, however much Mach might have emphasized these questions, if he had our present knowledge of the universe.

II. Mach's Principle and the Cosmological Constant [15]

The observational evidence for the structure of the universe, as it appears today, is summarized in several books, e.g. in G. C. McVittie's work [16]. It is customary to base cosmology on a principle, the so-called cosmological principle, which asserts that the universe is, apart from local irregularities, homogeneous and isotropic. One uses Einstein's equation

$$R_{nm} - \frac{1}{2} g_{nm} R^p_p + \lambda g_{nm} = -G\Theta_{nm} \qquad (7)$$

and inserts for Θ_{nm} the tensor due to a cosmic fluid in a classical description

$$\Theta_{nm} = (\rho + p) u_n u_m - g_{nm} p \qquad (8)$$

But rather than solve the equations (7) and (8), one appeals to the cosmological principle to justify a solution, the Robertson-Walker line element

$$(ds)^2 = dt^2 - \frac{R^2(t)}{(1-1/4\varepsilon r^2)^2} (dx^2 + dy^2 + dz^2) \qquad (9)$$

and expresses mass-density and pressure in terms of the parameters and functions in this solution. In (9), R(t) will be taken as a dimensionless function, and ε is then a quantity of $(\text{length})^{-2}$ and it can be shown that it can always be reduced to ±1 or 0.

The history surrounding the cosmological constant is well-known. As long as observations are not conclusive, it is safe to keep the constant for the sake of generality. But there does not seem to be any theoretical motivation for keeping it and some physicists prefer to omit it on these grounds. We will now show how the constant λ may be used to satisfy in part Mach's principle. In contradistinction to some models put forward and explained in the last section, we do not need to postulate a dynamical link between distant masses and local phenomena. A conceptional link is sufficient and also preferable on grounds of economy.

Classical field theory is based on the concept of

a Lagrangian which, incorporating the general theory of relativity, takes the form

$$\int d^4x(R^n_n g^{1/2} + aL(\phi;\phi_{,n};g_{nm})) = \text{Extremum} \qquad (10)$$

where L is the classical Lagrangian as a function of some fields ϕ, which may have various transformation properties. As we have explained in the first chapter, it is contrary to Mach's principle that the field equations make sense even in vacuum, where all fields ϕ are identically zero. The metric tensor will then be determined by the fundamental assumptions of the underlying space in the same way, as the number of dimensions of the space has to be assumed a priori. In any case, matter cannot be held responsible for the occurrence of inertial forces in some coordinate frames, since they would occur even in vacuum, merely as a consequence of the Riemannian nature of the manifold. To overcome this conceptional difficulty, we suggest supplementing the action principle (10) by a subsidiary condition, familiar from the calculus of variation,

$$\int d^4x(g^{1/2} - b\ M(\phi;\phi_{,n};\ g_{nm})) = 0 \qquad (11)$$

where b is a universal constant, and M a well defined function of the field variables. The meaning of this constraint is clear: the space-time extension of the universe should be equal to a given (positive definite) function of the physical variables, and in the "vacuum", the space-time loses its character of a metrical manifold. Before we discuss this assumption in any detail, we should add that the actual form of (11) was dictated mainly by the wish to preserve the general covariance of the theory. If one is willing to drop this requirement, other choices are possible, and the search for the correct form of the constraints is as much subject to empirical research as the search for the correct Lagrangian is. Although our form (11) complies with the rules of covariance, it necessitates a space-time with finite four-volume, in other words an oscillating universe, which may be at variance with observational cosmology.

We derive now the field equations. The Einstein equations are obtained by taking the Euler-Lagrange derivative with respect to the g_{nm}, and calling the Lagrange multiplier -2λ, we get

$$R_{nm} - 1/2\, g_{nm}\, R^p_p + \lambda g_{nm} = -a\, T_{nm} - 2\lambda b\, U_{nm} \tag{12}$$

where

$$T_{nm} = \partial L/\partial g^{nm} \tag{13}$$

and

$$U_{nm} = \partial M/\partial g^{nm} \tag{14}$$

On the other hand, the Euler-Lagrange derivative with respect to ϕ is

$$\delta(aL + 2\lambda bM)/\delta\phi = 0 \tag{15}$$

It is sometimes sufficient to study the Einstein equations (10) in the linear approximation in g_{nm}

$$g_{mn} = \delta_{mn} + \zeta_{mn} \tag{16}$$

and since

$$-R^n_n + 4\lambda = -k\, T^n_n - 2\lambda b\, U^n_n \tag{17}$$

we may write for (12)

$$R_{mn} + \lambda g_{mn} = -k(T_{mn} - 1/2\ T^p_p\ g_{mn}) + b(U_{mn} - 1/2 U^p_p g_{mn}) \tag{18}$$

and insert now (16) for R_{mn} and obtain as usual by a suitable coordinate transformation for the left hand side of (12)

$$\Box^2 \zeta + \lambda\zeta \tag{19}$$

where λ could be interpreted as the mass of the "gravitons". But the mass of the gravitons cannot be given arbitrarily, since it was introduced to satisfy the subsidiary condition.

It will now be necessary to make specific assumptions about the function M and the constants a,b. There is unfortunately no physical argument available, which would lead to a unique form of M, Mach's principle requires only that in the absence of some physical observables, i.e. fields, the function M should vanish. This may seem disappointing, if one had hoped that Mach's principle would in any way "explain" masses or interactions, a hope one could have harboured from the statement that "inertia should depend on the contents of the universe". But on the other hand, this broad formulation of Mach's principle, which is, just like the theory of relativity itself, a framework of theories, rather than a specific theory, has perhaps more hope of success.

Three general forms of the function M appear possible. One can identify M with the trace of the energy-momentum tensor, one can identify M with the mass term of those fields which carry a mass, and finally, one may make M equal to the interaction Lagrangian. Neither of these possibilities is dictated by any physical argument, but they do comply with the requirement of covariance; they are simple, e.g. they do not involve higher derivatives of the field variables, and they satisfy Mach's principle, if not in a strict sense, since M will not necessarily be positive definite with the above choices, at least in spirit.

The first possibility is in a sense the most complicated, because the energy-momentum tensor is not defined in closed form, since the right hand side of (12) contains the derivative of M with respect to g^{nm}. If we call the total energy momentum tensor, the source term in Einstein's equation, Θ_{nm}, we get the relation

$$T_{nm} + 2\lambda b \partial \Theta^{p}_{p} / \partial g^{nm} = \Theta_{nm} \tag{20}$$

thus a partial differential equation for the trace Θ^{n}_{n} which leads with its solution to a specific form of Θ_{nm}. Rejecting the idea that the most complicated possibility is also the most promising one, we turn now to the second possibility, assuming for L the Lagrangian of a scalar field

$$L = \frac{1}{2} g^{1/2} (\phi_{,n} \phi_{,m} g^{nm} - m^2 \phi^2) \tag{21}$$

where we have kept the mass term for the sake of generality, and set for M

$$M = \frac{1}{2} g^{1/2} m_1^2 \phi^2 \tag{22}$$

We obtain the field equations

$$(g^{1/2} g^{nm} \phi_{,n})_{,m} + g^{1/2} (m^2 - 2\lambda m_1^2 b/a)\phi = 0 \tag{23}$$

and similarly the Einstein equations

$$R_{mn} - 1/2\, g_{mn} R^{p}_{p} + \lambda g_{mn} =$$

$$= -a \left[\frac{1}{2}\phi_{,m} \phi_{,n} - \frac{1}{4}\phi_{,p} \phi_{,q} g^{pq} g_{mn} + \frac{1}{4}(m^2 - 2m_1^2 \lambda b/a)\phi^2 \cdot g_{mn}\right] \tag{24}$$

i.e. the usual expressions with an effective mass

$$m^2_{eff} = m^2 - 2m^2_1 \lambda b/a \tag{25}$$

We have here combined the two tensors (13) and (14) into a single energy-momentum tensor to permit a comparison with equation (7). The form implies that a is the constant of gravitation G, but this interpretation is far from self-evident. Certainly there is much ground to be covered between the fundamental energy-momentum tensors of quantum field theory, and their incorporation into general relativity, and the phenomenological tensor (8). If we nevertheless accept this identification of a with the gravitational constant, we note that for a sufficiently small G the mass becomes physically meaningless, assuming that the constant b is independent of G [17]. This is in qualitative agreement with the result of the corresponding limit in equation (6), if we assume there that the number of particles in the universe and its radius are kept fixed. Can one possibly explain the whole mass of a particle by the additional term in (25) and put m = 0 ? McVittie favours a negative cosmological constant, because it seems more in accord with the observational evidence in cosmology. However he favours also a negative ε [18], a space with infinite volume, which would make our constraint meaningless. Merely as an illustration, we will set m = 0, assume $a = Gh/c^3 = 10^{-66}$ cm^2 , and accept the upper limit for the cosmological constant 10^{-44} cm^{-2}. We get for bm^2_1 from the Compton wave length of a nucleon approximately 10^4 cm^2 . We should now calculate the proper vibrations of an oscillating universe from the combined wave equation (23) and the Einstein equation (24), but such a special solution would not be very useful. It is better to assume a realistic model, for instance that given by Hönl and Dehnen [19], which however these authors derived under the assumption of a vanishing cosmological constant.

The total four-volume of the universe is then about 10^{112} cm^4 , the duration of a cycle 10^{28} cm, and the greatest extension about 10^{27} cm. The present age of the universe corresponds to one fifth of the total cycle. The wave equation is of the type known as Hill's equation when suitable coordinates are employed. The lowest frequency will be about the inverse of the duration of a cycle, i.e. 10^{-28} cm, corresponding to a mass of 10^{-65} g. We note finally that the condition (11) allows us to estimate the average amplitude ϕ, and we get $\phi^2 = 2/bm_1^2 = 10^{-4}$ cm^{-2} . If we now try to identify the product of the mass of a proton with ϕ^2 as the mass density, which is at least dimensionally correct, we get the quite unreasonable result of 10^{22} cm^{-4}, while the observed mass density in the universe is at most 10^{-30} g/cm^3 , i.e. 10^7 cm^{-4} . These discouraging numerical excercises show at least, that the product bm_1^2 can be interpreted in two different ways, both very tentative. When multiplied with the constants λ and a, which were assumed to be known, we related bm_1^2 with the mass of particles. It is more natural, however, to consider the inverse of bm_1^2 as the eigenvalue of an "integrated number operator", a view which is suggested by the second part of our numerical illustration. In this case, the constant m will presumably have to be retained.

The third possibility of identifying M with various coupling terms of fields can be carried out in a similar manner. Here again, the coupling constants will be modified, and the question will arise whether the whole coupling can be ascribed to the subsidiary condition, or parts of it have to remain in the Lagrangian L.

If the function M depends on the metric tensor in the form

$$M = g^{1/2}\bar{M}(\phi;\phi_{,n}) \tag{26}$$

we can derive an interesting relation from the Einstein equation. The tensor U_{nm} in (14) will then have the form

$$U_{nm} = -\frac{1}{2} g^{1/2} g_{nm} \bar{M} \tag{27}$$

and in the integral over the whole universe of the trace of (12), the term in λ and the contribution of U_{nm} will just cancel, leaving

$$\int d^4x(R^n_n - a\, T^n_n)g^{1/2} = 0 \tag{28}$$

In the usual theory, the integrand itself vanishes. A general explicit equation for the constant λ can be found for any form of M, but the equation is of little use, except possibly for perturbation calculations.

There is another aspect of cosmology relevant to our subject. Fronsdal, Roman and Aghassi and recently Thirring and Nachtmann [20] have written on the subject of elementary particles in a curved space without including the general theory of relativity. Since they were primarily interested in generalizations of the ten-parameter Poincaré group, they restricted their studies to spaces of constant curvature, [21], which can be characterized by either

$$R_{nmpq} = c(g_{np}\, g_{mq} - g_{nq}\, g_{mp}) \tag{29}$$

hence

$$R_{nm} = -\,3c\, g_{nm} \quad , \tag{30}$$

or by considering hypersurfaces

$$x^2 + y^2 + z^2 - t^2 \pm u^2 = \pm r^2 \tag{31}$$

in the five-dimensional pseudo-Euclidian space with the metric

$$dS^2 = dt^2 - dx^2 - dy^2 - dz^2 \mp du^2 \qquad (32)$$

and one can show that either the upper or the lower sign must be taken in every case of ambiguity. Furthermore, the constants c and r are related by

$$c = \mp r^{-2} \qquad (33)$$

We do not need here the parameter representation of the hypersurfaces (31), which will give the de Sitter universe for the upper sign, nor will we need the generators of the Lie-algebra corresponding to the group of motion. It is obvious that the constant c can be related to a part of the source term in Einstein's equation, if these equations are accepted. We may set

$$\Theta_{nm} = \frac{1}{4} g_{nm} \Theta^p_p + (\Theta_{nm} - \frac{1}{4} g_{nm} \Theta^p_p) \qquad (34)$$

and identify

$$\pm 3/r^2 = -(\lambda + \frac{1}{4} G \langle \Theta^n_n \rangle) \qquad (35)$$

This equation, which will have to be assumed whenever a Riemannian space is approximated by a space of constant curvature, shows that there are two contributions to the curvature; not only the cosmological constant, but also a suitably averaged background energy. In the theory presented in this paper, the tensor Θ_{nm} will also involve the cosmological constant because of the tensor U_{mn}, and the additional contributions need not be small, compared with the first term in (35). However, these corrections, as well as the corrections to the orthogonality relations for the group of rotations in the five-dimensional space which result from the dependence of r on the fields, are clearly without practical significance at present.

References and Footnotes

1. see for instance H. Bondi, Cosmology, Cambridge University Press,1961, 2nd edition.
2. Ernst Mach, Die Mechanik in ihrer Entwicklung historisch-kritisch dargestellt, 2. verbesserte Auflage, Leipzig,F. A. Brockhaus;especially pp. 481, Mach emphasizes there that not only translational motion, but also rotational motion is relative.
3. A. Einstein, The Meaning of Relativity, Methuen & Co. London, 5th edition, 1951, pp. 95.
4. Einstein, ibid. p. 94. The occurrence of inertial forces in a frame in which the "stars" represented by a spherical mass-shell, are rotating, has been demonstrated by H. Thirring, Phys. Z. 19, 33,(1918); ibid. 22, 29(1921). The comments made by L. Bass and F. A. E. Pirani, Phil. Mag. 46, 850(1955), though important should not change the fundamental features of the conclusions. See also H. Hönl and A. W. Maue, Z. Phys. 114, 152 (1956) and Ch. Soergel-Fabricius, Z. Phys. 159, 541 (1960).
5. The action at a distance theories may serve to prove the point since the inclusion of all particles of the universe is contrary to the spirit of a physicist.
6. W. Thirring, Fortschritte d. Phys. 7, 79 (1959).
7. A. Einstein, loc. cit., p. 96, see also W. Davidson, Monthly No. Royal Astron. Soc. 117, 212 (1957).
8. loc. cit. p. 98.
9. D. W. Sciama, Monthly Not. Royal Astr. Soc. 113, 34, (1953); also "The Unity of the Universe", Doubleday Anchor, 1961.
10. Only exploratory papers have been published: O.

Bergmann, Am. J. Phys. 24, 38 (1956); Phys. Rev. 107, 1157 (1957). It was not proposed as a substitute for the general theory of relativity nor as interpreted by H. A. Buchdahl, Phys. Rev. 115, 1325 (1959) as a theory of "scalar charges".

11. I. Ozvath and E. Schücking, Nature 193, 1108(1962).
12. H. Dehnen and H. Hönl, Nature, 196, 362 (1962); H. Hönl and H. Dehnen, Zeits.f. Phys. 171, 178 (1963).
13. compare H. Hönl, Zeitschr.f. Naturf. 8a, 2 (1953).
14. The equation (6) follows easily from the H.Thirring's argument (Ref. 4). The Coriolis forces will depend on the masses and the radius of the rotating mass shell, and if these inertial forces should in fact be independent of these quantities, the equation (6) must be postulated.
15. A preliminary report was presented at the Autumn Meeting of the American Physical Society in Nashville, Tenn.,Bull. Am. Phys. Soc. 11, 819(1966). For other interpretations of the cosmological constant see F. M. Gomide, Nuovo Cim. 30, 672 (1963).
16. G. C. McVittie, General Relativity and Cosmology, The University of Illinois Press, Urbana,1965.
17. In a more sophisticated version of the theory the constant b may depend on a higher power of a, and the desire to have the masses vanish in the limit of vanishing gravitational constant could be satisfied assuming m = o (see page 301).
18. op. cit. p. 203.
19. H. Hönl and H. Dehnen, Z. f. Phys. 156, 382 (1959).
20. C. Fronsdal, Rev. Mod. Phys. 37, 221 (1965); P. Roman and J. J. Aghassi,Nuovo Cim. 152, 193 (1966); and other papers in preparation, W. Thirring, these proceedings; O. Nachtmann,to be published in Comm. Math. Phys.
21. L.P.Eisenhart,Riemannian Geometry,Princeton Univ. Press,(1963).

COULOMB FISSION BY VERY HEAVY IONS[†*]

By

E. GUTH et al.

Oak Ridge National Laboratory, Tenn., USA[**]

Abstract[***]

It is proposed that very heavy ions (including uranium) be utilized to indûce fission through the Coulomb interaction only. Because the projectile moves slowly, the process is expected to be nearly adiabatic (no intrinsic excitation). Dynamical model calculations have been performed at zero impact parameter to determine the threshold energy, cross section and fragment angular distribution. Differential cross sections of hundreds of millibarns are calculated, and fission fragments are found to emerge preferentially at 90° in the pair-frame. The calculations incorporate reasonable model data, but the equilibrium to saddle distance $\Delta\beta$ is unknown. In order for fission to occur below the Coulomb barrier, $\Delta\beta$ must be greater than .15 in the case of Cf, for example. Rotational and vibrational excitation is discussed for energies below the fission threshold. A primary objective

* to be published in Physical Review.
** Operated by Union Carbide Corporation for the U.S. Atomic Energy Commission.
*** Phys. Rev. Lett. 18(8), A9 (1967).
† Lecture given at the VI.Internationalen Universitätswochen für Kernphysik,Schladming,26 February-11 March 1967.

of the experiments would be to determine $\Delta\beta$ (and the shape of the energy-deformation curve). This would provide a severe test of various nuclear model theories.

LEPTONIC DECAYS OF THE KAON ACCORDING TO CURRENT ALGEBRA AND PCAC†

By

R. J. OAKES

CERN - Geneva

In this lecture I want to discuss the recent calculations of the leptonic decays of the K mesons based on the current commutation relations of Gell-Mann [1] and the assumption of a partially conserved axial vector current [2] (PCAC). For the most part this will be a review of the work of Callan and Treiman [3] and Weinberg [4], although some new results, obtained in collaboration with McNamee [5], will also be discussed. The presentation will largely follow an approach due to Bell [6] and will be rather elementary since this lecture is primarily intended for those who are not already familiar with the techniques.

The possible leptonic decays of the kaon are the following [7]

$$K_{\ell 2} : K \rightarrow \bar{\ell} + \nu_\ell$$

$$K_{\ell 3} : K \rightarrow \pi + \bar{\ell} + \nu_\ell$$

$$K_{\ell 4} : K \rightarrow \pi + \pi + \bar{\ell} + \nu_\ell$$

†Lecture given at the VI.Internationalen Universitätswochen f.Kernphysik,Schladming,26 February-11 March 1967

$K_{\ell 5}$: $K \to \pi + \pi + \pi + \bar{\ell} + \nu_\ell$

We shall discuss them all even though the latter process has never been observed to date. The K_{e5} mode is expected to be extremely rare but it is of some theoretical interest since, in calculating the $K_{\ell 5}$ amplitude, one encounters features that are not present in the other amplitudes.

Let us begin with the assumptions that underlie the calculations. Firstly, from the algebra proposed by Gell-Mann [1], we shall use the commutation relations

$$[Q_5^{(\alpha)}(t), V_\mu^{(\beta)}(\vec{x},t)] = -\sqrt{3}\begin{pmatrix} 8 & 8 & 8' \\ \alpha & \beta & \gamma \end{pmatrix} A_\mu^{(\gamma)}(\vec{x},t) \qquad (1)$$

and

$$[Q_5^{(\alpha)}(t), A_\mu^{(\beta)}(\vec{x},t)] = -\sqrt{3}\begin{pmatrix} 8 & 8 & 8' \\ \alpha & \beta & \gamma \end{pmatrix} V_\mu^{(\gamma)}(\vec{x},t) \qquad (2)$$

where $V_\mu^{(\alpha)}$ and $A_\mu^{(\alpha)}$ are the vector and axial vector currents, respectively, and $Q^{(\alpha)}$ and $Q_5^{(\alpha)}$ are the associated charges. That is,

$$Q^{(\alpha)}(t) = \int d^3x\, V_o^{(\alpha)}(\vec{x},t) \qquad (3)$$

and

$$Q_5^{(\alpha)}(t) = \int d^3x\, A_o^{(\alpha)}(\vec{x},t) \qquad (4)$$

We use, a,b,...n,m to indicate the SU(3) transformation properties and to label particles (antiparticles will be labelled $\bar{a}$, $\bar{b}$,...). Note that the normalization is such that the usual $\Delta S = \Delta Q = -1$, $\Delta\vec{I} = 1/2$ current is $(J_4 - iJ_5)_\mu = \sqrt{2}(V_\mu^{(K-)} - A_\mu^{(K-)})$. The SU(3) Clebsch-Gordan coefficients $\begin{pmatrix} 8 & 8 & 8' \\ \alpha & \beta & \gamma \end{pmatrix}$ are the standard ones to be found in the literature [8].

The second major ingredient is the hypothesis of a partially conserved axial vector current [2] (PCAC).

This can be stated in the form most convenient for our purposes by the relation

$$\sqrt{2}\,\partial^{\mu} A_{\mu}^{(\alpha)}(x) = C_{\alpha}\,\phi^{(\alpha)^{\dagger}}(x) \qquad (5)$$

where $\phi^{(\alpha)}(x)$ is the meson field and C_{α} is a real constant we shall determine below. The Hermitian conjugate field $\phi^{(\alpha)\dagger}$ enters in Eq. (5) since we follow the usual convention that, for example, the pion field $\phi^{(\pi+)}$ annihilates a π^{+} while the axial vector current $A_{\mu}^{(\pi+)}$ increases the isospin component. We have also adopted the usual phase convention for the fields and currents so that

$$\phi^{(\alpha)^{\dagger}} = \eta(\alpha)\,\phi^{(\bar{\alpha})} \qquad (6)$$

where

$$\eta(\alpha) = (-)^{(I_3+\frac{1}{2}Y)_{\alpha}} \qquad (7)$$

The normalization is chosen such that

$$<0|\phi^{(\beta)}(o)|\pi^{(\alpha)}(p)> = \delta_{\alpha\beta}(2p_o)^{-1/2} \qquad (8)$$

The PCAC constant C_{π} can be related to the pion lifetime. The pion decays entirely via the axial vector current and the relevant matrix element can be written in the form

$$<0|\sqrt{2}\,A_{\mu}^{(\beta)}(o)|\pi^{(\alpha)}(p)> = i\,\eta(\alpha)\,\delta_{\alpha\bar{\beta}}(2p_o)^{-1/2} m_{\pi} f_{\pi} p_{\mu} \qquad (9)$$

where f_{π} is a dimensionless form factor.

Then from Eqs. (5)-(9) one finds immediately that

$$C_{\pi} = m_{\pi}^{3}\, f_{\pi} \qquad (10)$$

From the commutation relations and the PCAC hypothesis one can derive the following very useful low energy theorem which relates a matrix element of the vector (axial vector) current to the axial vector (vector) current matrix for the process involving the emission of an additional zero frequency (unphysical) pion.

$$\lim_{p\to o}\sqrt{2p_o}\,\langle F\pi^{(\alpha)}(p)|\,\begin{matrix}V_\mu^{(\beta)}\\ A_\mu^{(\beta)}\end{matrix}\,|I\rangle =$$

$$= i\,\sqrt{6}(m_\pi f_\pi)^{-1}\,\eta(\alpha)\begin{pmatrix}8 & 8 & 8'\\ \alpha & \beta & \gamma\end{pmatrix}\langle F|\,\begin{matrix}A_\mu^{(\gamma)}\\ V_\mu^{(\gamma)}\end{matrix}\,|I\rangle \qquad (11)$$

To obtain Eq. (11) one first reduces in the pion and uses PCAC (Eq. (5)) to relate the pion field to the divergence of the axial vector current. Then, on partial integration and taking the limit $p\to o$, one finds the only surviving term can be evaluated using one of the commutators given above (Eqs. (1)-(2)). Before Eq. (11) can be applied to actual processes one must assume some relation between the physical matrix element where the pion is on its mass shell ($p^2=m_\pi^2$) and the matrix element in the soft pion limit ($p=o$). Following Bell [6] we shall make the rather optimistic assumption that in the extrapolation to vanishing pion momentum the matrix element is "as smooth as possible". That is, we shall take account explicitly of any contributions to a given matrix element that necessarily vary rapidly as the pion momentum vanishes, e.g., pole diagrams, and assume that the remainder varies so smoothly that any additional momentum dependence can be ignored. Of course, this principle of "Maximal Smoothness", which we adopt primarily because it is the simplest way to extrapolate, only assures us that we will avoid some manifest contradictions; nevertheless, a posteriori one can observe that it also leads to results consistent with the available experimental data.

As the first application we discuss the relation

between the $K_{\ell 3}$ and $K_{\ell 2}$ form factors obtained by Callan and Treiman [3]. The $K_{\ell 2}$ and $K_{\ell 3}$ form factors are defined by

$$<0|\sqrt{2}\ A_\mu^{(n)}(o)|K^{(m)}(k)> = i(2k_o)^{-1/2}\ m_K F_K\ k_\mu \tag{12}$$

and

$$<\pi^{(a)}(p)|\sqrt{2}\ V_\mu^{(n)}(o)|K^{(m)}(k)> =$$

$$= (4k_o p_o)^{-1/2}\left[F_+(k+p)_\mu + F_-(k-p)_\mu\right] \tag{13}$$

From isospin considerations we must have

$$F_K = \eta(n)\ \delta_{m\bar{n}}\ f_K \tag{14}$$

and

$$F_\pm = \sqrt{12}\ \begin{pmatrix} 8 & 8 & 8' \\ m & n & a \end{pmatrix}\ f_\pm \tag{15}$$

where f_K and $f_\pm$ depend only on the momenta. Applying the low energy theorem Eq. (11) to relate the $K_{\ell 3}$ amplitude, Eq. (13) to the $K_{\ell 2}$ amplitude, Eq. (12) gives

$$\sqrt{2}(f_+ + f_-) = \frac{m_K f_K}{m_\pi f_\pi} \tag{16}$$

which is the Callan-Treiman [3] result. In obtaining Eq. (16) we have neglected any momentum dependence of the form factors in accordance with the "Maximal Smoothness" principle discussed above, i.e., we can think of no contributions to the form factors that necessarily vary rapidly in the soft pion limit.

It is amusing to observe that had one applied the above considerations to the kaon rather than the pion one would have found the relation

$$\sqrt{2}\ (f_+ - f_-) = \frac{m_\pi f_\pi}{m_K f_K} \tag{17}$$

which, together with Eq. (16), implies that

$$(m_K f_K)^2/(m_\pi f_\pi)^2 = (1 + \xi)/(1 - \xi) \tag{18}$$

where $\xi = f_-/f_+$.

Hence, in the SU(3) symmetric limit $f_+ = 1/\sqrt{2}$ and $f_- = 0$, i.e., $\xi = 0$. If one includes the SU(3) breaking to first order by writing $(m_\pi f_\pi)/(m_K f_K) = 1 + x$ where x represents the deviation from symmetry, one finds $f_+ = 1/\sqrt{2} + O(x^2)$ and $f_- = O(x)$. This illustrates the Ademollo-Gatto [9] result, that SU(3) violating effects in f_+ are absent in lowest order [10].

Proceeding to the $K_{\ell 4}$ axial vector form factors we define [11]

$$<\pi^{(a)}(p_1)\pi^{(b)}(p_2)|\sqrt{2}\, A_\mu^{(n)}(o)|K^{(m)}(k)> =$$

$$= i\, m_K^{-1}\,(8k_o p_{10} p_{20})^{-1/2}[F_1(p_1+p_2)_\mu + F_2(p_1-p_2)_\mu +$$

$$+ F_3(k-p_1-p_2)_\mu] \tag{19}$$

where F_1-F_3 are invariant dimensionless functions of the momenta and isospins. Before applying the low energy theorem to relate the $K_{\ell 4}$ form factors to the $K_{\ell 3}$ form factors one must recognize that there is now a contribution that does indeed vary rapidly in the soft pion limit, namely, the K pole diagram shown in Fig. 1.

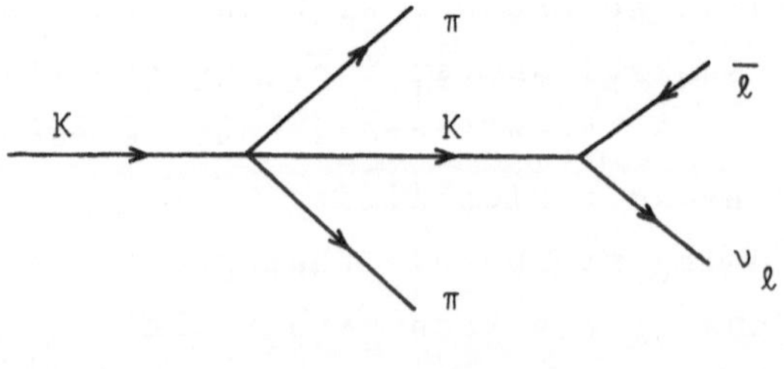

Fig. 1

K pole diagram in $K_{\ell 4}$

This contribution can be expressed in terms of the $K \to K\pi\pi$ amplitude, the kaon propagator, and the $K_{\ell 2}$ amplitude. Using Weinberg's [12] result for the $K \to K\pi\pi$ amplitude, one finds in the soft pion limit that the contribution is proportional to

$$\frac{k(p_1-p_2)}{k(p_1+p_2)} (k - p_1 - p_2)_\mu$$

which approaches quite different limits depending on which pion momentum vanishes. Consequently, it is impossible to neglect the momentum dependence of the K pole diagram, which contributes to F_3, when applying the low energy theorem to Eq. (19). We shall take the K pole contribution into account explicitly by defining

$$F_3 = f_3 + f \frac{k(p_1-p_2)}{k(p_1+p_2)} \tag{20}$$

Now, according to "Maximal Smoothness", we can neglect the momentum dependence of F_1, F_2, f_3, and f, there being no other rapidly varying contributions. Of course, these form factors still depend on the isospin, F_1 and f_3 being symmetric and F_2 and f being antisymmetric as required by Bose statistics. Now, applying the low energy theorem to the $K_{\ell 4}$ amplitude yields four relations among F_1, F_2, f_3 and f which are readily solved. The results, first given by Weinberg [4], are the following [13]:

$$F_1 = -\sqrt{2}\, \eta(a)\, \eta(m)\, \delta_{a\bar{b}}\, \delta_{m\bar{n}} (m_K f_+)/(m_\pi f_\pi) \tag{21}$$

$$F_2 = 6\, \sqrt{2} \begin{pmatrix} 8 & 8 & 8' \\ a & b & \ell \end{pmatrix} \begin{pmatrix} 8 & 8 & 8' \\ n & m & \ell \end{pmatrix} (m_K f_+)/(m_\pi f_\pi) \tag{22}$$

$$F_3 = \frac{1}{2} (1 + \xi) \left[F_1 + F_2 \frac{k(p_1-p_2)}{k(p_1+p_2)}\right] \tag{23}$$

Here $\xi = f_-/f_+$.

Finally I would like to mention a calculation of the $K_{\ell 5}$ decay amplitude done in collaboration with McNamee [5] at Stanford. If we define the $K_{\ell 5}$ vector form factors $G_1 \ldots G_4$ according to [14]

$$<\pi^{(a)}(p_1)\pi^{(b)}(p_2)\pi^{(c)}(p_3)|\sqrt{2}\, V_\mu^{(n)}(o)|K^{(m)}(k)> =$$

$$= m_K^{-2}(16k_o p_{10} p_{20} p_{30})^{-1/2} \times$$

$$\times \quad [G_1 p_{1_\mu} + G_2 p_{2\mu} + G_3 p_{3\mu} + G_4 (k - p_1 - p_2 - p_3)_\mu] \tag{24}$$

then, one can relate these to the $K_{\ell 4}$ axial vector form factors $F_1 \ldots F_3$ by means of the low energy theorem. However, a new complication arises in the extrapolation to zero pion momentum in that there are now rapidly varying contributions coming from the π pole diagram shown in Fig. 2(a) as well as from the K pole diagram shown in Fig. 2(b).

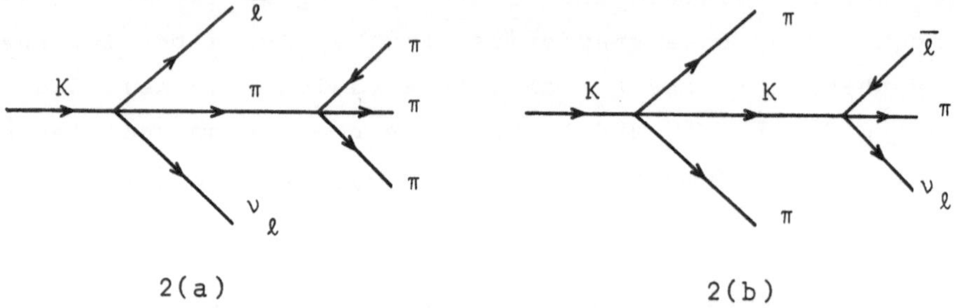

Fig. 2

We proceed as before by treating these contributions explicitly and neglecting the momentum dependence of the remainder. We define

$$G_1 p_{1\mu} + G_2 p_{2\mu} + G_3 p_{3\mu} + G_4 (k - p_1 - p_2 - p_3)_\mu \quad =$$

$$= g_1 p_{1\mu} + g_2 p_{2\mu} + g_3 p_{3\mu} + g_4 (k-p_1-p_2-p_3)_\mu + g_\pi \Pi_\mu + g_K K_\mu \qquad (25)$$

where Π_μ and K_μ represent the π pole and K pole contributions, respectively. These can be written out explicitly in terms of the $K \to K\pi\pi$, $\pi \to \pi\pi\pi$, and $K_{\ell 3}$ amplitudes in the soft pion limit, but are too lengthy to give here. Invoking "Maximal Smoothness" again we shall assume the form factors $g_1 \ldots g_K$ vary smoothly in the soft pion limit and treat them as constants depending only on the isospin involved. Then one applies the low energy theorem to relate $K_{\ell 5}$ to $K_{\ell 4}$, exactly as in the previous calculations, thereby obtaining rather lenghty general expressions for $G_1 \ldots G_4$ after some straightforward but tedious algebra. We give here only the result for the decay $K^+(k) \to \pi^+(p_1) + \pi^0(p_2) + \pi^-(p_3) + e^+ + \nu_e$ which is a typical case. (Complete results are contained in Ref. 5).) One finds the following:

$$G_1 = -2f_+ \left(\frac{m_K}{m_\pi f_\pi}\right)^2 \left[1 - 2\,\frac{(p_1+p_3)^2 - m_\pi^2}{(p_1+p_2+p_3)^2 - m_\pi^2}\right] \qquad (26)$$

$$G_2 = f_+ \left(\frac{m_K}{m_\pi f_\pi}\right)^2 \left[1 + 4\,\frac{(p_1+p_3)^2 - m_\pi^2}{(p_1+p_2+p_3)^2 - m_\pi^2} + \frac{k(p_1-p_3)}{k(p_1+p_3)}\right] \qquad (27)$$

$$G_3 = -2f_+ \left(\frac{m_K}{m_\pi f_\pi}\right)^2 \left[1 - 2\,\frac{(p_1+p_2)^2 - m_\pi^2}{(p_1+p_2+p_3)^2 - m_\pi^2} + \right.$$

$$\left. + \frac{k(p_1-p_2)}{k(p_1+p_2)}\right] \qquad (28)$$

$$G_4 = -\frac{1}{2}(f_+ + f_-)\left(\frac{m_K}{m_\pi f_\pi}\right)^2 \left[1 - 4\,\frac{(p_1+p_3)^2 - m_\pi^2}{(p_1+p_2+p_3)^2 - m_\pi^2} + \right.$$

$$\left. + 2\,\frac{k(p_1-p_2)}{k(p_1+p_2)} - \frac{k(p_1-p_3)}{k(p_1+p_3)}\right] \qquad (29)$$

Numerically the K_{e5} decay rates turn out to be $\sim 10^{-4}$-

- 10^{-3} sec^{-1} which are a little larger than previously thought [15], but still too small to be observed.

It has not been our purpose here to present a critical evaluation of these kaon decay calculations. However, in conclusion, we note that the above results agree with the data [7], where it exists, about as well as the experiments agree with each other. This is encouraging.

References

1. M. Gell-Mann, Phys. Rev. 125, 1064 (1962).
2. Y. Nambu, Phys. Rev. Lett. 4, 380 (1960); M. Gell-Mann and M. Lévy, Nuovo Cim. 16, 705 (1960).
3. C. G. Callan and S. B. Treiman, Phys. Rev. Lett. 16, 153 (1966).
4. S. Weinberg, Phys. Rev. Lett. 17, 336 (1966).
5. P. McNamee and R. J. Oakes, to be published.
6. J. S. Bell, Proceedings of the 1966 CERN School of Physics at Noordwijk-aan-Zee, CERN Yellow Report 66-29 (1966), and private communications.
7. For a recent review of the data see N. Cabibbo, Proceedings of the XIIIth International Conference on High Energy Physics, CERN Report TH. 711.
8. J. J. de Swart, Revs. Modern Phys. 35, 916 (1963); P. McNamee and F. Chilton, Revs.Modern Phys. 36, 1005 (1964).
9. M. Ademollo and R. Gatto, Phys. Rev. Lett. 13, 264 (1964).
10. I am indebted to J. S. Bell for this observation.
11. There is also a vector term $\sim F_+ \varepsilon_{\mu\nu\rho\sigma} k^\nu p_1^\rho p_2^\sigma$ that we shall not discuss. It cannot contribute in the soft pion limit and presumably is negligible in the physical region owing to the centrifugal barrier.
12. S. Weinberg, Phys. Rev. Lett. 17, 616 (1966).
13. In Eqs. (21) - (23) we have corrected a sign error in Eq. (24) of Ref. 4).

14. The axial vector form factors are presumably unimportant. See Ref. 11).
15. V.A.Kolkunov and I. V. Lyagin, Zhur.Eksp.i Teoret.Fiz. 45, 2009 (1963); English translation: Sov. Phys. JETP 18, 1379 (1964).

VECTOR MESONS AND THE ELECTROMAGNETIC INTERACTIONS OF HADRONS†

By

H. JOOS

DESY, Hamburg

Introduction

Since about three years information on the interaction of photons and electrons with nucleons is coming in from the electron accelerators in the Multi-GeV-region. I would like to discuss the question how this has improved our understanding of the electromagnetic (e.m.) interaction of the hadrons, i.e. of the strongly interacting particles. My procedure will be to consider several experimental results under the uniform aspect of a single model, the vector meson dominance model (VDM) of the e.m. interaction of hadrons [1].

In order to get a first impression of the phenomenological situation I give a compilation of the total cross-sections for the photoproduction of the different particles and resonances in table I. These are rough average values in two energy intervals. For precise information I must refer to the quoted literature [2].

† Lecture given at the VI. Internationalen Universitätswochen f.Kernphysik,Schladming,26 February-11 March 1967.

$\gamma+p\rightarrow$	2-3,5 GeV μb	3,5-5,5 GeV μb	Literature
$\pi^{+}n$	3		2a,b,e
$\pi^{o}p$	1 - 2		2a,c
$\pi^{-}N^{*++}$	6	3	2d,f
$\rho^{o}p$	17	16	2d,f,g
ωp	5	3	2d,f
ϕp	0,5	0,4	2d,f
$\rho^{-}N^{*++}$	2,0	0,6	2d,f
fp	2,5	0,6	2d
$X^{o}p$	0,5	0,2	2d
$\pi^{+}\pi^{+}\pi^{-}\pi^{-}p$		4	2h

Table I

Total cross-sections for photo-production processes.

The dominant feature, particularly at high energies, is the large cross-section for ρ^{o}-production. From this fact one may get the idea that the ρ^{o}-meson is essential for the coupling of the e.m. field to the hadrons. One may remember that the ρ^{o} was in a way predicted by W. R. Frazer and J. R. Fulco [3] from the e.m. form factors of the nucleons. Now the following is the simplest form of a relation between the e.m. current $j_{\mu}(x)$ of the hadrons and the phenomenological field $\rho^{o}_{\mu}(x)$ of the ρ^{o}-meson or more general with the fields of the vector mesons ρ^{o},ω and ϕ :

$$j_{\mu}(x) = - \left[\frac{m_{\rho}^{2}}{2\gamma_{\rho}} \rho^{o}_{\mu}(x) + \frac{m_{\omega}^{2}}{2\gamma_{\omega}} \omega_{\mu}(x) + \frac{m_{\phi}^{2}}{2\gamma_{\phi}} \phi_{\mu}(x)\right] \qquad (1)$$

m_V, $V = \rho^{o}, \omega, \phi$, denotes the masses of the vector mesons. The constants γ_V are partly determined by the symmetry properties of the e.m. current and the vector mesons. This simple "Ansatz" (1) gives a successful starting point for the understanding of the general features of the e.m. interactions of hadrons as I want to show in this talk. Eq. (1) is the essence of the VDM.

In the first part I want to discuss some theoretical ideas on the meaning of eq. (1) and in the following section I shall discuss the application of this model to the problems of the e.m. form factors, the leptonic and the radiative decays of the mesons, the photoproduction of mesons etc. I would like to emphasize that it is my aim to give a general view point on the problems of the e.m. interactions of hadrons and not a demonstration of a numerically best fitting model with a theoretically most detailed description for a special case.

I. Some Theoretical Ideas

I want to discuss different theoretical aspects of the meaning of eq. (1).

a) The e.m. current and the interpolating fields of the vector mesons

We first consider eq, (1) under the most general point of view, namely in the framework of general field theory. In order to simplify this discussion, we treat the vector mesons as stable particles. M. Gell-Mann and F. Zachariasen [1] have indicated how one may understand this as an approximation. In general field theory one considers the different quantized fields only under the aspects of relativistic covariance, causality respectively locality, particle description etc., but one does not assume special field equations. It is an important result [4] of the scattering theory in this framework that any general field A(x) which carries the quantum numbers of a particle β may be used to describe the scattering processes of this particle. This result is relevant to our problem. We shall formulate this statement more explicitly but without technical details:

Let $A(x)$ be a local covariant field operator in a Hilbert space $\mathcal{H}$ with the vacuum state $|0\rangle = \Omega$ and a one β-particle state $|\beta\rangle$, which is a normalizable eigenstate of the mass operator $P_\mu P^\mu$ with the eigenvalue m_β^2, and which corresponds to a normalized one particle wave function $e^\beta(x)$ such that

$$\langle 0|A(x)|\beta\rangle = C_\beta . e^\beta(x) \neq 0 \tag{2}$$

then there exist in $\mathcal{H}$ incoming and outgoing free fields $\beta^{in}(x)$, $\beta^{out}(x)$:

$$(\Box - m_\beta^2)\beta^{ex}(x) = 0 \quad , \quad [\beta^{ex}(x), \beta^{ex}(x')] = i\Delta(x-x';m_\beta^2)$$

$$ex = in,\ out \tag{3}$$

which are related to $A(x)$ by the asymptotic condition [5]:

$$C_\beta^{-1} A(x) = \beta^{\substack{in\\ out}}(x) + \int dx'\ \Delta_{\substack{ret\\ av}}(x-x', m_\beta^2) j_\beta(x)$$

$$j_\beta(x) = C_\beta^{-1}(\Box - m_\beta^2)\ A(x) \tag{4}$$

The scattering matrix elements for incoming and outgoing particles with given momentum may be expressed with help of the creation and annihilation operators $b^+_{ex}(k)$, $b_{ex}(k)$ of β-particles

$$\beta^{ex}(x) = (2\pi)^{-3/2}\int \frac{d^3k}{2k_o}(e^{-ikx} b_{ex}(k) + e^{ikx} b^+_{ex}(k))$$

$$k_o = \sqrt{\vec{k}^2 + m_\beta^2} \tag{5}$$

and the operators $c^+_{ex}(k)$, $c_{ex}(k')$ ... of the other particles. These may be calculated in the same manner from A(x) or from another field C(x) which is relatively local to A(x) : $[A(x),C(x')] = 0$ for $(x-x')^2 < 0$. The S-matrix element for a process $\beta+\gamma\rightarrow\beta'+\gamma'+\ldots$ has the form:

$$\langle\beta',\gamma',\ldots|S|\beta,\gamma\rangle = (b^+_{out}(k')c^+_{out}(p')\ldots\Omega, b^+_{in}(k)\, b^+_{in}(k)c^+_{in}(p)\Omega) \quad (6)$$

We have suppressed in this description the superficial problems related to the spin and internal degrees of freedom. In physics one assumes always for an interacting field that it satisfies condition (2) when this assumption does not contradict the conservation laws, i.e. if field and particle carry the same quantum number. A field A(x) which satisfies condition (2) for a certain particle β and which is normalized such that $C_\beta = 1$ is called an interpolating field for β. There are many different interpolating fields which lead to the same causal S-matrix. We don't have principles in general field theory, which select special interpolating fields. Special fields are selected by field equations.

From these considerations we gain the interpretation of equation (1). Since we don't have any established theory with field equations for the vector meson field, we may try the zero component of the iso-spin current $j^{(o)}_\mu(x)$, the hyper-charge current $j^Y_\mu(x)$ and the current of the baryonic charge $j^B_\mu(x)$ as interpolating fields of the vector particles. We have to take care of the quantum numbers, we have to use therefore $j^{(o)}_\mu(x)$ as the interpolating field of the neutral isovector particle ρ^o, we define

$$\rho^o_\mu(x) = -\frac{2\gamma_\rho}{m^2_\rho}\, j^{(o)}_\mu(x) \;, \text{ i.e. } \langle 0|j_\mu(o)|\rho\rangle = = -\frac{m^2_\rho}{2\gamma_\rho}\, e_\mu(\rho) \quad (7)$$

It is in the spirit of our treatment of the vector mesons as stable particles to assume that the real ω and ϕ carry different quantum numbers. SU(3)-symmetry and some of its breaking schemes ("ω-ϕ-mixing") provide such quantum numbers. We assume that ϕ carries the quantum numbers of

$$j_\mu^\phi(x) = -\cos\theta\ j_\mu^Y(x) + \sin\theta j_\mu^B(x)$$

and ω the quantum numbers of

$$j_\mu^\omega(x) = \sin\theta\ j_\mu^Y(x) + \cos\theta\ j_\mu^B(x)\ .$$

So we define as the corresponding interpolating fields

$$\phi_\mu(x) = \frac{C_\phi}{m_\phi^2}(\cos\theta j_\mu^Y(x) - \sin\theta j_\mu^B(x))\ ,$$

$$\text{i.e.}\quad \langle 0|j_\mu^\phi(o)|\phi\rangle = -\frac{m_\phi^2}{C_\phi}\, e_\mu(\phi)$$

$$\omega_\mu(x) = -\frac{C_\omega}{m_\omega^2}(\sin\theta j_\mu^Y(x) + \cos\theta j_\mu^B(x))\ ,$$

$$\text{i.e.}\quad \langle 0|j_\mu^\omega(o)|\omega\rangle = -\frac{m_\omega^2}{C_\omega}\, e_\mu(\omega) \tag{8}$$

and we note

$$\langle 0|\omega_\mu(x)|\phi\rangle = \langle 0|\phi_\mu(x)|\omega\rangle = 0\ . \tag{9}$$

We don't have in this approximation virtual transitions between ω and ϕ [6]. The use of conserved currents as interpolating fields has the consequence that we describe the vector particles by transversal fields

$$\frac{\partial}{\partial x^\mu}\rho_\mu^o(x) = \frac{\partial}{\partial x^\mu}\omega_\mu(x) = \frac{\partial}{\partial x^\mu}\phi_\mu(x) = 0 \tag{10}$$

Now it follows from the Gell-Mann Nishijima formula for the e.m. current

$$j_\mu^{em}(x) = j_\mu^{(o)}(x) + \frac{1}{2} j_\mu^Y(x) =$$

$$= - (\frac{m_\rho^2}{2\gamma_\rho} \rho_\mu^o(x) - \frac{m_\phi^2}{2C_\phi} \cos\theta \phi_\mu(x) + \frac{m_\omega^2}{2C_\omega} \sin\theta \omega_\mu(x)) =$$

$$= - (\frac{m_\rho^2}{2\gamma_\rho} \rho_\mu^o(x) + \frac{m_\phi^2}{2\gamma_\phi} \phi_\mu(x) + \frac{m_\omega^2}{2\gamma_\omega} \omega_\mu(x)) \quad (11)$$

We have thus an interpretation of ansatz eq. (1): "The different parts of the e.m. and baryonic currents with respect to internal hadron-symmetry may be used as interpolating field of the vector mesons". SU(3)-symmetry implies $\sqrt{3}\gamma_\rho = C_\phi = C_\omega$. The mixing angle of SU(6): $\cos\theta = \sqrt{2/3}$ results in:

$$\gamma_\rho : \gamma_\omega : \gamma_\phi = 1 : 3 : - 3/\sqrt{2} \quad (12)$$

What are the practical consequences of this interpretation? From the asymptotic condition we derive the T-matrix element for a process V+A → B+C:

$$<p_B;p_C|T|k_V,s_V;p_A> = (2\pi)^{5/2} <p_B,p_C|K_\mu^V(o)|p_A> \times$$
$$\times e_\mu^{s_V}(k_V)\delta(p_B+p_C-k_V-p_A) \quad (13)$$

with $K_\mu^V(x) = (\Box - m_V^2)V_\mu(x)$.
$e_\mu^s(k)$ is the polarization vector of the vector meson; we have suppressed the spin indices of the particles A,B,C.

Similarly we get for the T-matrix element of a photo-production process γ+A → B+C

$$<p_B,p_C|T|k,s;p_A> =$$

$$= e(2\pi)^{5/2} <p_B,p_C|j_\mu(o)|p_A> a_\mu^s(k)\delta(p_B+p_C-k-p_A) \quad (14)$$

$\Box A_\mu(x) = e\, j_\mu(x)$ is the e.m. current, $A_\mu(x)$ the vector potential of the e.m. field, $a_\mu^s(k)$ the polarization vector of the photon with momentum k and polarization s. With a look at our definitions (7) and (8) the equations (13) and (14) seem to differ only by a factor

$$- \frac{e}{2\gamma_V} \frac{m_V^2}{(k^2-m_V^2)}$$

in the different parts of the e.m. current.
But there is an important difference. In eq. (13) we have

$$k^2 = (p_B + p_C - p_A)^2 = m_V^2$$

whereas in eq. (14) we have

$$k^2 = (p_B + p_C - p_A)^2 = 0$$

Now we may try the assumption that the matrix element of $K_\mu(o)$ is slowly varying in k^2 for fixed m_A^2, m_B^2, m_C^2 and fixed $t_{AC} = (p_A-p_C)^2$ or fixed $t_{AB} = (p_A-p_B)^2$ or under any other kinematical condition. One necessary condition that such an extrapolation procedure makes sense is that one does not come close to threshold-type or other kinds of kinematical singularities by varying k^2 from m_V^2 to 0. As a consequence of the transversality condition (10) $k_\mu K_\mu(o) = 0$, the matrix element of the longitudinal current $e_\mu^{long}(k)K_\mu(o)$ vanishes for $k^2 = 0$, only the transversal helicity amplitudes have limits different from zero. Having talked about all those problems, I neglect the variation of

$$\langle p_B, p_C | K_\mu^V(o) | p_A \rangle$$

in k^2 in the calculation of the cross-sections for the processes $\gamma + A \rightarrow B + C$ resp. $V + A \rightarrow B + C$ and I get the nice result connecting the differential or total photoproduction cross-section $\sigma(\gamma+A \to B+C)$ with corresponding cross-sections $\sigma_{tr}(V+A \to B+C)$ for transversally polarized V-mesons:

$$\sigma(\gamma+A \to B+C) \simeq \frac{\alpha\pi}{\gamma_V^2}\, \sigma_{tr}\, (V+A \to B+C) \qquad (15)$$

in the special case, when only one part of the e.m. current (compare eq. (11)) contributes. In the more general case one has to add the different contributions and one must discuss possible interference effects. One may derive similar formulas for the production with virtual photons, i.e. for electro-production.

The relation between the V-meson current and the e.m. current

$$\langle p,\alpha | j_\mu^{(V)}(o) | p',\beta \rangle =$$

$$= \frac{-e}{2\gamma_V} \frac{m_V^2}{((p-p')^2 - m_V^2)} \langle p,\alpha | K_\mu^V(o) | p',\beta \rangle \qquad (16)$$

may be described formally in the language of Feynman diagrams by a two-line vertex

with a vertex part $- \frac{em_V^2}{2\gamma_V} g_{\mu\nu}$

and an outgoing γ-line.

In the next chapter we shall show that eq. (15) and eq. (16) may be applied very successfully to many processes. Therefore it is worthwile to think on further justification of the prescribed procedures. From first principles one may hopefully derive some analyticity in k^2 under favourable conditions [7]. In the following sections we mention shortly some other physical philosophies.

b) Vector dominance in the dispersion theoretic approach

The dispersion theoretic approach to the VDM gives some further insight in the assumptions which are hidden in our extrapolation procedure. We discuss this shortly in the case of the e.m. form factor of the pion:

$$<p|j_\mu(o)|p'> = \frac{1}{(2\pi)^3}(p+p')_\mu F_\pi((p-p')^2) \tag{17}$$

This is the classical case for which M. Gell-Mann first discussed the VDM. Only the isovector part $j_\mu^{(o)}(x)$ contributes in (17). A dispersion relation can be proved for $F_\pi(k^2)$, we assume it in the unsubtracted form

$$F_\pi(k^2) = \frac{1}{\pi}\int_{4m_\pi^2}^{\infty} \frac{A(t)}{t-k^2-i\varepsilon}\, dt \tag{18}$$

with the absorptive part

$$A(t) = 2\pi \sum_n^{\text{all states}} \frac{p'_\mu}{4m_\pi^2-t} <o|j_\mu^{(o)}(o)|n><n|J_\pi(o)|p'> \times$$
$$\times \quad \delta(t-m_n^2) \tag{19}$$

The extrapolation procedure described in (a) would give the result

$$<p|j_\mu(o)|p'> = \frac{-1}{2\gamma_\rho}\frac{m_\rho^2}{(p-p')^2-m_\rho^2} <p|K_\mu(o)|p'> =$$
$$= \frac{1}{(2\pi)^3}\frac{g_{\rho\pi\pi}}{2\gamma_\rho}\frac{m_\rho^2}{m_\rho^2-(p-p')^2}(p_\mu+p'_\mu) \tag{20}$$

where we consider the "slowly varying" ρ^o-form factor of the pion approximated by the $(\pi-\pi-\rho^o)$-coupling constant $g_{\rho\pi\pi}$. The same result we get in the dispersion theoretic approach if we take only the term with the ρ^o in the intermediate state:

$$A_{\rho^o}(t) = 2\pi \int \frac{d\vec{p}_\rho}{2p_o^\rho}\sum_s \frac{p'_\mu}{4m_\pi^2-t} \times$$
$$\times \quad <0|j_\mu^{(o)}(o)|p_\rho,s><p_\rho,s|J_\pi(o)|p'>\delta(t-m_\rho^2) =$$

$$= \pi \cdot g_{\rho\pi\pi} \frac{m_\rho^2}{2\gamma_\rho} \delta(t-m_\rho^2) \tag{21}$$

The error we make by neglecting the k^2 dependence of the ρ^o-form factor of the pion can be seen from the dispersion relation [1c]

$$F_\pi(k^2) = \frac{g_{\rho\pi\pi}}{2\gamma_\rho} \frac{m_\rho^2}{m_\rho^2-k^2} + \frac{1}{\pi} \int_{4m_\pi^2}^{\infty} \frac{A(t)-A_{\rho^o}(t)}{t - k^2 - i\varepsilon} dt \tag{22}$$

It is of course possible that with our special choice of the interpolating ρ^o-field our approximation procedure is bad and we have to take into account the ρ^o-form factor.

c) A reference to Lagrangian field theory

One may consider the relation (1) between the vector meson fields and the e.m. current as part of field equations which are derived from a Lagrangian. The idea that the vector mesons are coupled "universally" to conserved currents was proposed in 1960 by J.J.Sakurai [1a]. He assumed an effective Lagrangian for the ρ-coupling of the type

$$\mathcal{L}_{eff} = f_{\rho NN} \vec{\rho}_\mu \bar{N} \gamma_\mu \frac{\vec{\tau}}{2} N + f_{\rho\pi\pi} \vec{\rho}_\mu (\vec{\pi} \times \partial_\mu \vec{\pi}) + \ldots =$$
$$= \sum_{i=1}^{3} f_\rho \rho_\mu^{(i)} j_\mu^{(i)} \tag{23}$$

which would lead to field equations

$$(\Box - m_\rho^2)\rho_\mu^{(i)}(x) = f_\rho j^{(i)} \tag{24}$$

and the "universality" condition [8]

$$f_\rho^2/4\pi = f_{\rho NN}^2/4\pi = f_{\rho\pi\pi}^2/4\pi \qquad (= 2{,}4-2{,}8) \tag{25}$$

In this field equation the isovector current appears as source of the ρ^o-meson field and not as the interpolating field of the ρ^o-meson. In order to decide if Sakurai's model is different from VDM one must discuss the Lagrangian field theory with its full content, in particular the renormalization of the theory. Such an analysis was performed in a recent paper by Kroll, T. D. Lee and B. Zumino [1]. They describe a Lagrangian field theory in which the relation

$$\rho_\mu^o(x) = \frac{2\gamma_\rho}{m_\rho^2} j_\mu^{(o)}(x)$$

holds for the renormalized fields-those are the interpolating fields- and for which eq. (24) holds for the unrenormalized fields under the condition that the unrenormalized mass $m_o^2 = \infty$. For further details we must refer to this paper. The well-known problems of field equations for strongly interacting particles unfortunately restrict the use of field equations to general questions.

II. A Review of Applications of the Vector Dominance Model

The following considerations are to demonstrate that the vector meson dominance model gives a good overall picture of the e.m. interactions of the hadrons. We use as a short hand the graphical description.

a) The leptonic decays of the vector mesons [9]

An immediate consequence of the virtual V-γ-transition is the decay of the vector mesons in electron-respectively muon-pairs. We may calculate the partial width according to the Feynman graph fig. 1 .

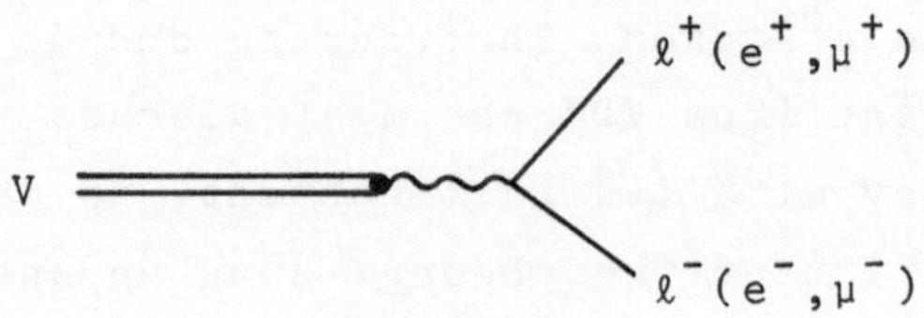

Fig. 1

The result is

$$\Gamma(V\to\ell^+ + \ell^-) = \frac{\alpha^2}{12}\left(\frac{4\pi}{\gamma_V^2}\right) m_V \left(1 - \frac{4m_\ell^2}{m_V^2}\right)^{1/2}\left(1+\frac{2m_\ell^2}{m_V^2}\right) \simeq$$

$$\simeq \frac{\alpha^2}{12}\left(\frac{4\pi}{\gamma_V^2}\right) m_V \qquad (26)$$

The branching ratio

$$R^\ell = \Gamma(V\to\ell^+ + \ell^-)/\Gamma(\rho_o\to\pi^+ + \pi^-)$$

is approximately the same for e-decay and μ-decay. The following values one can find in the literature:

ρ^o:	$R^e = 5.10^{-5}$	(R.A.Zdanis et al., Phys. Rev. Lett. 14, 721 (1965))
	$R^\mu = 4,3.10^{-5}$	(Wehmann et al., Phys. Rev. Lett. 17, 1113 (1966))
	$R^\mu = (4,4\pm 2).10^{-5}$	(Weinstein, 13.Int.Conference on High Energy Physics, Berkeley 1966)
	$R^e = 3,9.10^{-5}$	(M.A.Azimov, Dubna preprint, E1-3148 (1967))

An average value of $R = 4,5.10^{-5}$ would give $\gamma_\rho^2/4\pi$ the value $\gamma_\rho^2/4\pi = 0,54$ ($M_{\rho_o} = 770$ MeV, $\Gamma(\rho_o\to 2\pi) = 140$ MeV, Rosenfeld Jan. 1967).

There are no reliable measurements on the ω and φ decay.

b) The form factors of the pseudoscalar mesons [10]

The simple VDM allows to calculate the pion form factor from the graph of fig. 2

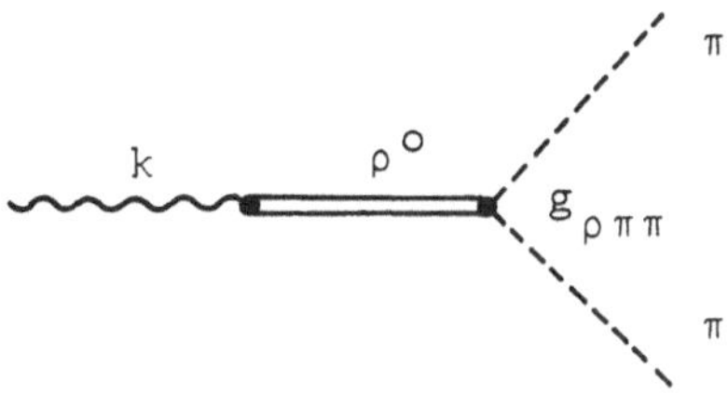

Fig. 2

with the result

$$F_\pi(k^2) = \frac{g_{\rho\pi\pi}}{2\gamma_\rho} \frac{m_\rho^2}{m_\rho^2 - k^2} \tag{27}$$

(see eq. (20)).

The coupling constant $g_{\rho\pi\pi}$ can be calculated from the width of the decay

$$\Gamma(\rho\to\pi^+ + \pi^-) = \frac{g_{\rho\pi\pi}^2}{4\pi} \frac{m_\rho}{12} \left(1 - \left(\frac{2m_\pi}{m_\rho}\right)^2\right)^{3/2} \tag{28}$$

From the normalization condition F(o) = 1 we conclude

$$\gamma_\rho^2/4\pi = \frac{1}{4} \frac{g_{\rho\pi\pi}^2}{4\pi} = 0{,}67$$

The charge radius of the pion is according to this model

$$r_\pi = \sqrt{6 \left| \partial F/\partial k^2 \right|_{k^2=o}} = \frac{\sqrt{6}}{m_\rho} \simeq 0{,}6 \text{ f} \tag{29}$$

The best experimental value [11] comes from the discussion of the process $e+p\to e+n+\pi^+$ and has the value $r_\pi = 0{,}7\pm$ $\pm 0{,}1$ f which is in reasonable agreement with our value.

We may apply the same reasoning to the K^+-form factor. There all V contribute. Unfortunately there are no experiments of the reaction

$$e + p \to e + K^+ + \Lambda \quad \text{or} \quad e + p \to e + K^+ + \Sigma^0$$

from which one could get some idea of r_{K^+}. For the discussion of the $\overline{K^0}$, K^0-form factor we refer to Kroll et al. [1].

c) Electromagnetic decays of mesons [12]

The decays $\omega \to \pi^0+\gamma$ and $\pi^0 \to \gamma+\gamma$ are two classical subjects of the application of the vector-dominance model. The result of the calculations corresponding to the graphs fig. 3 is according to the calculations of Gell-Mann, Sharp, Wagner [1]:

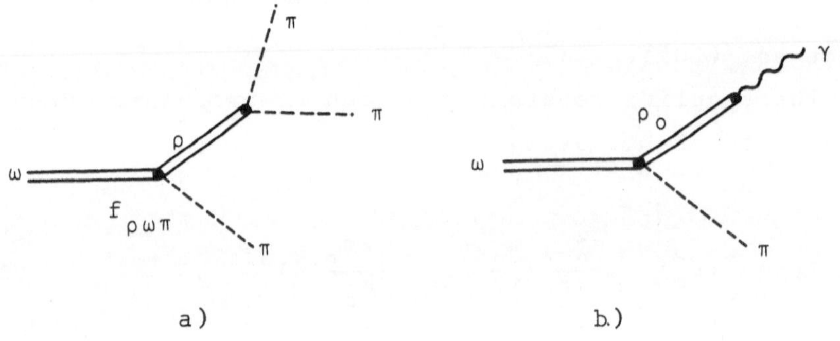

Fig. 3

a)

$$\Gamma(\omega\to 3\pi) = \left(\frac{g^2_{\rho\pi\pi}}{16\pi}\right)\left(\frac{f^2_{\rho\omega\pi}}{4\pi}\right)\frac{(m_\omega - 3m_\pi)^4}{(m_\rho^2 - 4m_\pi^2)^2}\,\frac{m_\omega m_\pi^2}{\sqrt{27}}\,W(m_\omega)$$

$$W(m_\omega) \simeq 3{,}56$$

b)

$$\Gamma(\omega\to\pi^0+\gamma) = \frac{\alpha}{96}\left(\frac{f^2_{\rho\omega\pi}}{4\pi}\right)\left(\frac{\gamma_\rho^2}{4\pi}\right)^{-1}\frac{(m_\omega^2 - m_\pi^2)^3}{m_\omega^3}$$

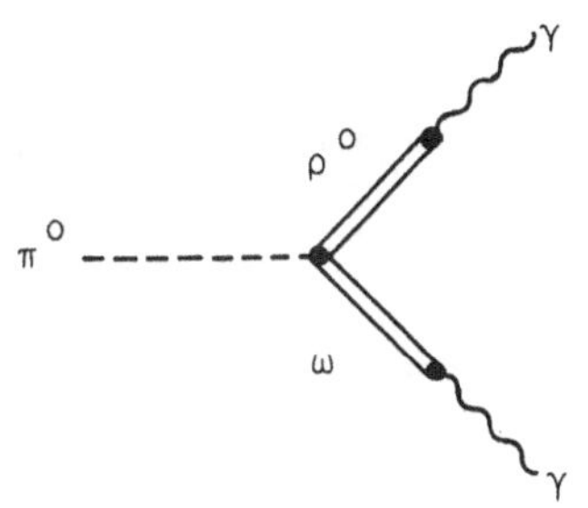

Fig. 3c

c)

$$\Gamma(\pi^{o}\to 2\gamma) = \frac{\alpha^2}{192}\left(\frac{\gamma_\rho^2}{4\pi}\right)^{-1}\left(\frac{\gamma_\omega^2}{4\pi}\right)^{-1}\left(\frac{f^2_{\rho\omega\pi}}{4\pi}\right) m_\pi^3 \qquad (30)$$

We may get a value of $f^2_{\rho\omega\pi}/4\pi$ from the decay $\omega\to 3\pi$ according to the graph a. The values for

$\Gamma(\omega\to 3\pi) = 10{,}7$ MeV

$\Gamma(\omega\to\pi^{o}+\gamma) = 1{,}16$ MeV

$\Gamma(\pi^{o}\to 2\gamma) = 7{,}4$ eV

from Rosenfelds table (Jan. 67) result in

$\gamma_\rho^2/4\pi \approx 0{,}66$, $\gamma^2_{\ \omega}/4\pi \approx 3{,}6$

The analysis [2d] of the peripheral part of the reaction $\gamma+p\to\omega+p$ leads to a somewhat smaller

$\Gamma(\omega\to\pi^{o}+\gamma) = 0{,}67$ MeV

In this type of analysis there are certainly many incertainties with respect to the treatment of the final state interactions and the pionic form factors. There are new resonances $A_1(1080)$ and $A_2(1300)$ which mainly decay into $A\to\rho^{o}+\pi$. The vector dominance model suggests

$$\Gamma(A^{\pm}\to\pi^{\pm}+\gamma) = \frac{\alpha\pi}{\gamma_\rho^2} X_{tr} \times$$

$$\times\Gamma(A^{\pm}\to\rho^{o}+\pi^{\pm})\quad\left[\frac{(m_A^2-(m_\rho+m_\pi)^2)(m_A^2-(m_\rho-m_\pi)^2)}{(m_A^2-m_\pi^2)^2}\right]^{1/2} \tag{31}$$

X_{tr} = fraction of A's decaying into transverse polarized ρ^o. It seems not impossible to measure this width with help of the reaction (fig.4)

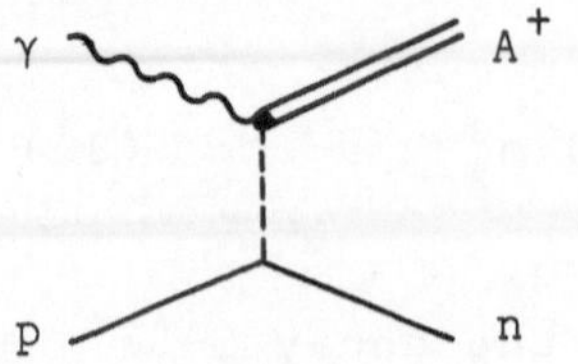

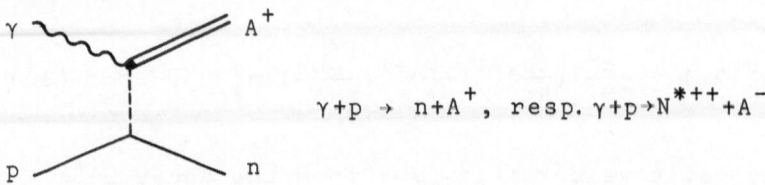

$\gamma+p \to n+A^+$, resp. $\gamma+p\to N^{*++}+A^-$

Fig. 4

Similar considerations may be applied to the strange resonances $K_A(1320)$ and $K_A(1420)$ which have decay modes $K_A \to K + \rho^o$, (ω).

d) The form factor of the nucleons

Now we apply our simple minded VDM to the analysis of the nucleon form factors. Similarly to the case of the meson form factors, we get expressions of the Clementel-Villi type for the electric and magnetic (Sachs-)form factors G_e^I, G_m^I for the two isospins I = 0, 1:

$$G^o_{e,m}(q^2) = \frac{a^\omega_{e,m}}{1-q^2/m^2_\omega} + \frac{a^\phi_{e,m}}{1-q^2/m^2_\phi}$$

q^2 < scattering

$$G^1_{e,m}(q^2) = \frac{a^\rho_{e,m}}{1-q^2/m^2_\rho} + \frac{a^{\rho'}_{e,m}}{1-q^2/m^2_{\rho'}} \tag{32}$$

where the constants $a^V_{e,m}$ are composed of the (V-N-$\bar{N}$)-coupling constants $g^{VN}_{e,m}$ and the γ_V : $a^V_{e,m} = g^{VN}_{e,m}/2\gamma_V$. Unfortunately the simple VDM with ρ^o, ω and ϕ is not in agreement with the experiments in the iso-vector case. Here we would have only the pole from the ρ^o . But the experiments indicate very strongly that the form factors decrease faster than $1/|q^2|$ for $q^2 \to -\infty$. Further we get from the ρ^o-mass a charge radius which is too small. One must add in the I = 1 channel an additional term, as we did in (32).

Table II contains the values of the constants $a^V_{e,m}$ from two recent fits [13] , [14] to the experimental form factors [15]. The values of the a^V are not yet very consistent. One reason for this is, that $m_{\rho'}$ is badly determined by the fitting procedure.

	Fit I [13]	Fit II [14]
a^ω_e	1,18	1,29
a^ω_m	1,05	1,20
a^ϕ_e	-0,68	-0,81
a^ϕ_m	-0,61	-0,84
a^ρ_e	2,27	1,20
a^ρ_m	7,37	5,33
$a^{\rho'}_e$	-1,77	-0,70
$a^{\rho'}_m$	- 5,01	-3,04
$m_{\rho'}$	900 MeV	980 MeV
m_ρ	760 MeV	750 MeV

Table II.

Constants of a dipole fit to the nucleon form factor.

What is the meaning of the additional part in the isovector part? We can think of three possibilities:

a) there is a not yet discovered vector meson which contributes.

b) it represents a general background term which is not related to a special intermediate state of the absorptive part.

c) the term may be interpreted as ρ^o- form factor (this is a physical version of (b)):

$$G^1_{e,m}(q^2) = \frac{F^{e,m}_{\rho_o}(q^2)}{1 - q^2/m^2_\rho} \tag{33}$$

This last version was recently discussed by T. Massam and A.Zichichi [16]. In the isoscalar channel we have the two vector-mesons ϕ and ω and therefore enough parameters to fit the form factors - if we don't use symmetry arguments for the coupling constants. But using a ρ^o-form factor makes it natural to use also a ω,ϕ-form factor.

I think the situation in the theory of the form factors of the nucleon is not yet transparent enough to get a reliable determination of γ_V. Only as an example I adopt position (a) and determine

$$\gamma_{\rho_o} \quad : a^\rho_m = g^{\rho N}_m / 2\gamma_\rho$$

with help of the magnetic coupling constant $g^{\rho N}_m$ [17]:

$$\gamma^2_\rho / 4\pi \simeq 0,19 \; ... \; 0,36$$

e)Photoproduction of vector mesons [18]

We may apply formula (15) to the process (fig. 5) $\gamma + p \to V + p$:

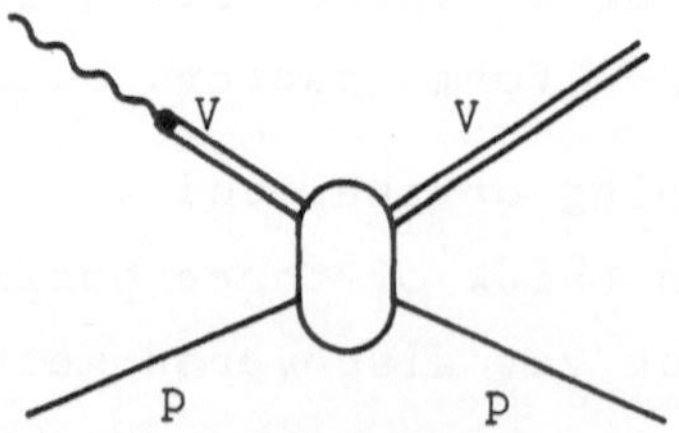

Fig. 5

In the diffraction region, i.e. at high energies and small momentum transfer, we may assume that the scattering amplitude is equal for the different helicity states, formula (15) then becomes

$$\sigma(\gamma+p\to V+p) = \frac{\alpha}{4}\left(\frac{\gamma_V^2}{4\pi}\right)^{-1}\sigma(V+p\to V+p) \qquad (34)$$

The VDM explains therefore the frequency of the ρ^o in photoproduction as diffraction effect. In order to calculate diffraction cross-section for V+p→V+p we use sum rules for the total cross-sections $\sigma_T(V+p)$ which follow from Lipkin's quark model [19]

$$\sigma_T(\rho^o p) = \sigma_T(\omega p) = \frac{1}{2}(\sigma_T(\pi^- p) + \sigma_T(\pi^+ p))$$

$$\sigma_T(\phi p) = 2\sigma_T(K^+ p) + \sigma_T(\pi^- p) - 2\sigma_T(\pi^+ p) \qquad (35)$$

For the differential cross-section in the diffraction region we make the ansatz

$$\frac{d\sigma}{dt} = \left(\frac{d\sigma}{dt}\right)_{t=o} e^{at}$$

We assume that the main contribution to the elastic cross-section comes from the diffraction part, then we get with help of the optical theorem:

$$\sigma_{diff}(V+p\to V+p) = \frac{1}{16\pi a}\sigma_T^2(Vp) \qquad (36)$$

A reasonable value of a is 8 GeV^{-2}. This is a rough value from $\pi^{\pm}p$ scattering (a = 7.6 GeV^{-2}) and it describes the angular distribution of the ρ^o-photoproduction at lab. energies K = 3,5-5 GeV. We put a equal for ρ^o,ω and ϕ. This may bring some errors in our calculation! $\sigma(\gamma+p\to\rho_o+p)$ has the features of a diffraction cross-section, whereas below 6 GeV peripheral processes contribute to the ω-production. In [2d] we find a numerical analysis, which contributes 1,8 μb to the diffraction mechanism from a total of 3μb. Little is known about the ϕ-production. As the diffraction cross-section varies little with energy, we may calculate with eqs.(34), (35), (36) the coupling constants γ_ρ, γ_ω, γ_ϕ from the data of $\pi^\pm p$, $K^\pm p$ reactions which we took from the compilation of Lipkin's paper. The result is in table III. Most remarkable is the explanation of the small ϕ cross-section by the smallness of the Kp-cross-section compared to the πp-cross-section.

	Lab.mom. GeV	ρ^o	ω	ϕ	units
$\sigma_T(Vp)$	6	27,3	27,3	10,1	mb
$\sigma_{diff}(\gamma+p\to p+V)$	3,5<k<5,5	16	1,8(3,0)	0,4	μb
$\sigma_{diff}(Vp)$	6	4,8	4,8	0,65	mb
$\gamma_V^2/4\pi$		0,55	4,9(2,9)	3	

Table III

Theoretical and experimental values of cross-sections for V - p reactions

f) The total cross-section for the production of hadrons by photons [18]

We may apply the reasoning of the preceding section to get an estimate of the total hadronic photon cross-section $\sigma_T(\gamma p)$. The formula

$$\sigma_T(\gamma p) = \frac{\alpha}{4}\left[\left(\frac{\gamma_\rho^2}{4\pi}\right)^{-1}\sigma_T(\rho_o p) + \left(\frac{\gamma_\omega^2}{4\pi}\right)^{-1}\sigma_T(\omega p) + \left(\frac{\gamma_\phi^2}{4\pi}\right)^{-1}\sigma_T(\phi p)\right] \qquad (37)$$

neglects interference effects between contributions from different V-mesons as well as a possible dependence of $\sigma_T(Vp)$ on the helicities.

With

$$\gamma_\rho^2/4\pi = 0{,}55 \ , \qquad \gamma_\omega^2/4\pi = 4{,}9 \ , \qquad \gamma_\phi^2/4\pi = 3$$

(compare tables III and V) we get for $k = 6$ GeV

$$\sigma_T(\gamma p) = 106\,\mu b$$

There are no experimental data available.

g) The photoproduction of pions [20]

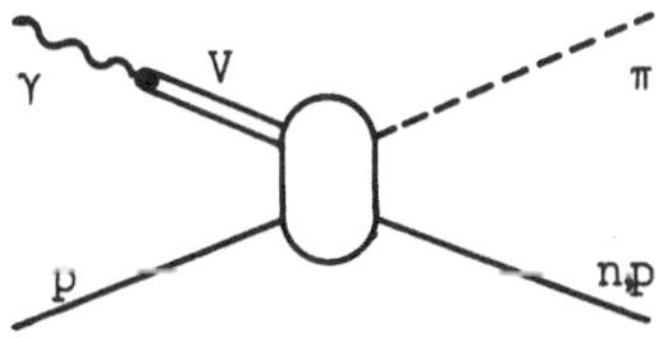

Fig. 6

Formula (15) relates the photoproduction of pions to the production of pions by vector mesons and this is related by time reversal to the V-production by pions. We make an isospin analysis, assume that the main contribution comes from the ρ_o-part, $(1/\gamma_\rho^2 = 9 \cdot \frac{1}{\gamma_\omega^2} = \frac{9}{2} \cdot \frac{1}{\gamma_\phi^2})$

$$\frac{d\sigma}{dt}(\gamma + p \rightarrow \pi^+ + n) = \frac{1}{2}(1 - X_{oo}) \frac{|\vec{p}_\pi|}{|\vec{p}_{\rho_o}|} \frac{\alpha . \pi}{\gamma_\rho^2} \frac{d\sigma}{dt}(\pi^- p \rightarrow \rho^o n) \qquad (38)$$

$|\vec{p}_\pi|$ and $|\vec{p}_{\rho_o}|$ are the momenta of π^- and ρ_o in the cms. $(1-X_{oo})$ describes the projection on the transversal components of the ρ_o. P. Schmüser [21] compared cross-sections for $\gamma+p\to\pi^++n$ with values $\sigma^{theor.}$ calculated with help of eq. (38) and experimental data for the reaction $\pi^-+p\to\rho^o+n$ [22]. For the total cross-section he gets with $\gamma_\rho^2/4\pi = 0,57$ (compare table V):

k_{lab} GeV/c	$\sigma^{exp}(\pi^-+p\to\rho^o+n)$	$\sigma^{theor}(\gamma+p\to\pi^++n)$ $(1-X_{oo})=0,7$	$\sigma^{exp}(\gamma+p\to\pi^++n)$
1,7	~ 2 mb [22a]	3,2 μb	6 μb
2,75	1,1 mb [22b]	1,4 μb	2,9 μb
3	1,6 mb [22c]	2,1 μb	2,8 μb

Table IV

VDM and π^+ - photoproduction

The slope of the differential cross-section agrees reasonably well at k = 1,7 GeV/c. At the higher energies the slope of

$$\frac{d\sigma^{theor}}{dt}(\gamma+p\to\pi^++n)$$

is considerably steeper than the experimental one. Regarding all the experimental incertainties in $\sigma^{exp}(\pi^-+p\to\rho^o+n)$, in X_{oo} and its t-dependence etc. we believe that this discussion of the process $\gamma+p\to n+\pi^+$ with VDM is quite promising. We expect even better results at higher energies, where the difference between the ρ^o-mass and the γ-mass = 0 should be less important.

h) A remark on the excitation of nucleon isobars by photons

The VDM suggests that nucleon isobars N^* which have a large decay width in the channel $N^*\to N+\rho^o$ may be easily excited by γ-rays:

$$\gamma + N \to N^{*} \to \begin{cases} N + \rho^{o} \\ N + \pi \end{cases} \quad \text{etc.}$$

Unfortunately the decay channel $N^{*} \to N + \rho^{o}$ of the higher isobars is experimentally not yet studied, and the photoproduction data are not yet analyzed under this aspect. I believe $N^{*}(1920)$ and $N^{*}(2120)$ would be interesting candidates.

i) Conclusion

In order to give an impression of the consistency of the VDM I have put together all results on $\gamma_V^2/4\pi$ in table V. I think with the VDM we have improved our qualitative understanding of the e.m. interactions of the hadrons.

	$\gamma_\rho^2/4\pi$	$\gamma_\omega^2/4\pi$	$\gamma_\phi^2/4\pi$
$V \to \ell^{+} + \ell^{-}$	0,54		
π-formfactor	0,67		
$\omega \to \gamma + \pi$, $\pi^{o} \to 2\gamma$	0,66	3,6	
N-formfactor	0,19-0,36		
$\gamma + p \to V + p$	0,55	4,9	3
$\gamma + p \to \pi^{-} + n$ (at 3 GeV)	0,76		

Table V

Different determinations of the γ-V-coupling constants

During the preparation of this talk I had valuable discussions on scattering theory with Prof. Dr. R. Haag and on experimental data with Drs. E. Lohrmann, P. Schmüser and G. Wolf. I would like to thank my colleagues.

References

1. Some of the papers on the VDM are:
   a) J. J. Sakurai, Ann. Phys. 11, 1 (1960);
   b) M. Gell-Mann and F. Zachariasen, Phys. Rev. 124,953 (1961);
   c) M. Gell-Mann, Phys. Rev. 125, 1067 (1962);
   d) J. J. Sakurai, Proceedings of the International School of Physics "Enrico Fermi" Course XXVI(1963);
   e) Y. Nambu and J. J. Sakurai, Phys. Rev. Lett. 8, 79 (1962);
   f) M. Gell-Mann, D. Sharp and W. G. Wagner, Phys. Rev. Lett. 8, 261 (1962);
   g) R. F. Dashen and D. H. Sharp, Phys. Rev. 133, B1585 (1964);
   h) L. Stodolsky, Phys. Rev. 134, B1099 (1964);
   i) G. Barton and B. G. Smith, Nuovo Cim. 36, 436 (1965);
   k) M. Roos and L. Stodolsky, Phys. Rev. 149, 1172 (1966);
   l) D. S. Beder, Phys. Rev. 149, 1203 (1966);
   m) N. M. Kroll, T. D.Lee, B. Zumino, Preprint: Neutral vector mesons and the hadronic electromagnetic current (preprint 1967).

   The treatment of the e.m. interactions of hadrons in the quark model is closely related to the VDM; see for example:
   W. Thirring, Proc. of the V. Int. Universitätswochen für Kernphysik, Schladming 1966, p. 294.
2. a) L. S. Osborne, Proc. Int. Symp. on Electron and Photon Interactions, Hamburg 1965 and literature quoted there;

b) W. Bertram et al., DESY-Bericht,Phys. Rev. Lett. 17, 1027 (1966);

c) M. Braunschweig et al., Phys. Lett. 22, 105 (1966);

d) Aachen-Berlin-Bonn-Hamburg-Heidelberg-München Collaboration DESY-Bericht 66/32; Nuovo Cim. 47,675 (1967), and Nuovo Cim. 48, 262 (1967); Phys. Lett. 23, 707 (1966) and literature quoted there;

e) D. O. Caldwell et al., CEA-preprint;

f) Cambridge-Bubble-Chamber-Group, Proc. Int.Symp. on Electron and Photon Interactions,Hamburg 1965, Vol. II; Phys. Rev. 146, 994 (1966); Phys. Rev., to be published;

g) W. Lancerotti et al.,Phys. Rev. Lett. 15, 210(1965);

h) E. Lohrmann, private communication.

3. W. R. Frazer and J. R. Fulco, Phys. Rev. 117, 1603 (1960); Y. Nambu, Phys. Rev. 106, 1366 (1957);
4. R. Haag, Phys. Rev. 112, 669 (1958); K. Nishijima, Phys. Rev. 111, 995 (1958); W. Zimmermann, Nuovo Cim. 10, 597 (1958);
5. G. Källen, Helv. Phys. Acta 25, 417 (1952); H. Lehmann, K. Symanzik, W. Zimmermann, Nuovo Cim. 1, 425 (1955); Nuovo Cim. 6, 319 (1957);
6. Compare M. Ross and L. Stodolski, Phys. Rev. Lett. 17, 563 (1966);
7. Y. S. Jin, "Analyticity of Forward Scattering Amplitude in External Mass", Brown Univ. preprint (1966), submitted to Nuovo Cim.;
8. J. J. Sakurai, Phys. Rev. Lett. 19, 1021 (1966).
9. 1 c, e, f, g.
10. 1 c, 3.
11. C. W. Akerlot et al.,Phys.Rev. Lett. 16, 147 (1966), Dubna Conference (1967).
12. 1 f; S. Hori et al., Phys. Lett. 1, 81 (1962).
13. R. Wilson, Proc. Int. Symp. on Electron and Photon Interactions, Vol. I, Hamburg 1965.
14. H.Hultschig, private communication; I thank Dr. Hult-

schig for a discussion on the subject.
15. W. Albrecht et al., Phys. Rev. Lett. 17, 1193(1966); W. Bartel et al., Phys. Rev. Lett. 17, 608 (1966).
16. T. Massam and A. Zichichi, Nuovo Cim. 43,1137 (1966).
17. G. Köpp, G. Kramer, Phys. Lett. 19, 593 (1965) and literature quoted there.
18. S. M. Berman, S. D. Drell, Phys. Rev. 133, B791(1964) and F. Buceella and M. Colocci, Phys.Lett. 24B, 61 (1967); P. G. O. Freund, Nuovo Cim. 44,A411 (1966); preprint 1966; M. Ross, L. Stodolsky, Phys. Rev. 149, 1172 (1966); H. Joos, Phys. Lett. 24B, 104 (1967); K. Kajantie, I. S. Trefil, Phys. Lett. 24B, 106 (1967).
19. H. J. Lipkin, Phys. Rev. Lett. 16, 1015 (1966).
20. D. S. Beder, Phys. Rev. 149, 1203 (1966).
21. P. Schmüser (Diss. Hamburg 1967).
22. a) D.D.Allen et al., Phys. Rev. Lett. 17, 53 (1966);
b) Saclay-Orsay-Bari-Bologna-Collab., Nuovo Cim. 35, 713 (1965);
c) W. Selove et al., Phys. Rev. Lett. 9, 272 (1962).

SUMMARY - First Week†

By

H. PIETSCHMANN

Institut für Theoretische Physik,
Universität Wien, Austria

There is an end to everything and - according to tradition - the last talk at a conference is the summary; in a short form, it is supposed to review all the lectures, partly for those who preferred to breathe the wonderful fresh air of Schladming's skiing slopes while others listened to the talks in the nice but less sunny Stadtsaal. This year, the summary is split into two parts, one for each week, and it is the privilege of the first summarizer to heartily thank Professor Urban and his able staff for the very fine organization of the school and its relaxed atmosphere.

The first week of this year's "Internationale Universitätswochen für Kernphysik" comprised two seminars and four lecture series. The seminars were given by F. Rohrlich on "Magnetic Monopoles" and by O. Bergmann on "Mach's Principle".

Among the lectures, I will first comment on L. Browns talk on "Meson-Decays". They were dedicated to the students in the audience in accord with the principle that the first week should be of introductory nature. The technical tools which are necessary to actually compute a decay rate were presented in a very pedagogical way and in the name of all younger colleagues who might feel

† given at the VI. Internationalen Universitätswochen f.Kernphysik,Schladming,26 February-11 March 1967.

a bit lost on a level where phase space integrations and the like are too trivial even to be mentioned, I would like to thank Prof. Brown very cordially for the time he has taken to prepare these very fine lectures.

The subject of all four lecture series lies in the field of strong interactions. In Quantum Electrodynamics, we are fortunate to possess a theory that allows for a computation of measurable quantities from the coupling constant and the masses of the participating particles alone. The theory of strong interactions is far from this happy state. Therefore, the only possibility left is to correlate various measured quantities to each other. In this field of phenomenology, great progress has been made over the last decade and a satisfactory calculation has to agree with experiments to within 3-5% which is about one order of magnitude better than one generally requires in nuclear physics, for example.

Dispersion Relations provided the main tool for many years in the field of strong interaction phenomenology. Only recently, they have been complemented or paralleled by new methods. From V. de Alfaro we heard the recent progress that has been achieved with the method of superconvergence conditions. It is worthwhile noting that both in his as well as in J.J.J.Kokkedee's lectures, Regge poles played a major role. This is very comforting because I think they were thrown overboard with no real good reason. We should remember, that the Regge-pole model was able to predict a particle with <u>all</u> its properties, that is with all quantum numbers of the vacuum except that its spin was predicted to be 2 and the mass slightly above 1 GeV. The discovery of the f^{o} with exactly these properties proved that there was at least some truth in the model. Recall that the success of SU(3) is analogously based upon the prediction of the Ω^{-} which was subsequently found. The difference, however, being that we have so far no experimental proof that its spin-parity is really $3/2^+$.

Part of the lectures of J.J.J.Kokkedee were concerned with the application of the quark model to high energy scattering. Here, we definitely deviate from pure phenomenological relations and try to understand hadrons on a dynamical basis. However the success of the model seems to be too good to be comforting for we would expect deviations between experiments and the rather crude model calculations of the order of those in nuclear physics, i.e. about 20-50%. A serious discrepancy has so far been found only in one case, namely the decay

$$\eta \to \pi^o + 2\gamma$$

which has a quark model branching ratio of the order 10^{-3} whereas present measurements yield order unity. The discrepancy is generally blamed on the measurements; thus a careful investigation of this decay should be encouraged.

M. Nauenberg's lectures treated the most important of the new tools in hadron phenomenology, current algebra. With the example of the Mathur-Okubo-Pandit-Callan-Treiman Relation he showed, how relations are obtained from essentially 3 ingredients: The reduction formula (RF), current commutation relations (CCR) and partial conservation of the axial vector current (PCAC). Symbolically, one can express this in the following way:

$$\langle a\pi | J_\mu | b\rangle \xrightarrow{RF} \int \langle a | [\phi, J_\mu] | b\rangle$$

$$\langle a | [\phi, J_\mu] | b\rangle \xrightarrow{PCAC} \langle a | [\partial_\mu j_\mu, J_\mu] | b\rangle$$

$$\langle a | [j_\mu, J_\mu] | b\rangle \xrightarrow{CCR} \langle a | J_\mu | b\rangle$$

With this chain, one relates e.g. the decays $K^+ \to \to \ell^+ + \nu_\ell$ and $K^+ \to \pi^o + \ell^+ + \nu_\ell$. There was always the feeling that PCAC is the only really crucial assumption in the game and that one really tests PCAC rather than CCR. To support this, I would like to remind you that excellent relations on strong interaction quantities have been derived by S. Adler from PCAC alone (the so-called "consistency conditions"). We have here learned, that current commutation relations can in fact be derived from certain assumptions on the derivative of the currents. If the ordinary PCAC condition

$$\partial_\mu j^5_\mu = c\phi_\pi$$

is complemented on the right hand side by terms proportional to the electromagnetic or weak coupling constant,current commutation relations actually follow. This is very important in itself and moreover sheds light on the famous difficulty with Schwinger terms. But I am now improperly beginning to talk about things which actually were discussed in the second week. Before I conclude, I would like to comment on a question which has been raised after one of M. Nauenberg's lectues. I do not recall it word by word, but in essence it was the following: "Do you know, whether your time dependent generators exist at all?". It is certainly true that one has troubles in mathematically rigorously defining these objects and it is necessary to devote much effort to these questions. On the other hand, this should not prevent phenomenologists from using these "undefined" objects because otherwise progress would come to an abrupt halt. Quantum electrodynamics, for example, would not be what it is now if questions of mathematical rigour had stopped the computation of Feynman diagrams at an early stage.

But now it is time for me to yield to Professor Källén for the summary of the second week.

SUMMARY - Second Week †

By

G. KÄLLEN

Department of Theoretical Physics
University of Lund, Sweden

A large part of the lectures during the second week has been concerned with two very general frame works of modern theoretical physics, viz. S-matrix theory and field theory. Perhaps it may be appropriate to start with a general remark about these two approaches.

In S-matrix theory attention is concentrated on a particular object, the S-matrix, which is alleged to describe every observable consequence of a given theory. Characteristic for the S-matrix is that only quantities on the mass shell enter. From the mathematical point of view this means that fields at finite times are not used but only asymptotic fields which are obtained when the time in the formalism goes to plus or minus infinity. From the philosophical point of view it might not be out of place to remark that this is a mathematical construction which can never really be realized exactly. Very long ago ($t = -\infty$) the observer was not even born and in the very distant future ($t = +\infty$) he will certainly be dead. All real experi-

† given at the VI. Internationalen Universitätswochen für Kernphysik, Schladming, 26 February-11 March 1967.

ments in the laboratory have to be performed during a finite time interval and, therefore, correspond to a situation which cannot be exclusively described in terms of on-the-mass-shell objects but which involve finite times and off-shell quantities (fields). Therefore, the S-matrix as it is mathematically defined contains quantities which are, in principle, completely unobservable.

After this very provocative statement let us at once admit that this is only a question of principle and that a time interval of, say, 10^{-10} sec is usually an extremely good approximation of an infinite time interval. Therefore, S-matrix theory is actually concerned with quantities which are very close to those objects which one does observe in practice. Consequently, it is certainly a subject which is very well worth studying.

This year, Polkinghorne has in a series of lectures given a very clear summary of the work which has been done in Cambridge, England, about the singularity structure of the S-matrix. In this investigation one exploits the consequences of certain general assumptions about the S-matrix, viz. unitarity, Lorentz invariance, analyticity, crossing symmetry etc. The last property allows one to relate transition amplitudes for two processes which differ only by some particles in the final state in one of the reactions being removed and replaced by corresponding antiparticles in the initial state. As the physical variables entering in the different "channels" do not have the same numerical values the assumption about crossing symmetry is useful only in connection with some prescription about how to continue the S-matrix elements from one region to another in the space of the external variables. This is done with the aid of the analyticity assumption. Roughly speaking, the assumption here is that every element of the S-matrix is assumed analytic until proved singular. This approach should be contrasted to other arguments by other groups where one takes some pains to assure that the S-matrix (or similar objects) are really analytic in certain domains and assumes that no analyticity exists except where it can be proved. In the Cambridge approach the opposite is done. However, also in Cambridge the S-matrix cannot be arbitrarily analytic as the unitarity relation with necessity generates singularities at various reaction thresholds etc. Polkinghorne has described the technique used by the Cambridge group to identify and - to a certain extent - to describe the singularities made necessary by the unitarity equation. The procedure is somewhat complicated and, I believe, iterative. The final result appears not to be available in the literature today but Polkinghorne has announced that a series of papers is forthcoming where all important consequences are worked out in the physical region. For the moment the result of the first few iterations has been published. Usually, the singularities obtained can be recognized from perturbation theory calculations.

A different point of view more closely related to orthodox field theory and using also off-the-mass-shell concepts has been described by Rohrlich. He has given a summary of the work done by him and his students (especially Pugh) about a new formulation of the equations of motion in a field theory - especially in electrodynamics. The usual equations of motion in the Heisenberg picture involving infinite renormalization constants have been replaced by more complicated but mathematically better defined equations involving rather sophisticated concepts from the theory of generalized functions. It appears that the usual renormalization prescriptions are effectively replaced by certain "boundary conditions" of the integro-differential equations used in this formulation. Further, no explicit Lagrangian is used but at a certain stage a "Born term" (or zero order vertex part) is introduced in the calculation. This explicit form of the Born term distinguishes the various interactions from each other. The technique is claimed to be able to handle also non renormalizable theories to a limited extent.

Even if some of the mathematical details especially the form of the equation of motion for the free fields used in non renormalizable theories as well as the exact form of the boundary conditions replacing the renormalization prescriptions appear to need further clarification and perhaps modifications I am personally willing to believe that it is possible to construct a consistent formalism along the lines indicated by Rohrlich. However, I must confess that, apart possibly from some points of mathematical rigour, the usefulness of this approach appears to me not to have been demonstrated. Even if it is literally correct that no infinite quantity is explicitly encountered during any step of the calculation, the infinities are still there in the theory even if they are better hidden than in the usual method. Personally, I do not believe in "the renormalization philosophy" i.e. in the idea that quantities like electromagnetic self energies etc. are, in principle, unobservable. As a counter example I should like to quote the mass differences between particles inside the same isospin multiplet like the π-mesons, the K-mesons, the various baryon multiplets etc. These mass differences are physically observable and - at least probably - due to electromagnetic effects. Therefore, it is a physically important problem to calculate these mass differences from a basic theory and I feel it is not enough to be able to write the formalism in such a way that the masses are put in in the form of a boundary condition. As is obvious for those of you who have listened to the lectures this last week Rohrlich has, at this point, an opinion which is very different from my own.

No conference today is complete without at least one lecture about group theory. In Schladming this year we have had the pleasure to listen to a series of talks by Budini about a new way of using group theory to calculate mass spectra. The general idea appears to be to replace the conventional idea about an interaction by the alternative assumption that the physical world can be completely described by invariance properties. The mass operator in a relativistic theory or the Hamiltonian in a non relativistic theory is supposed to be invariant under a rather large group, not necessarily compact. This assumption means that the mass operator can be written as a function of the Casimir operators of the large group. Further, one is supposed to use only a particular representation of the large group which is a direct sum of all the physically important representations of e.g. the Lorentz group in the relativistic case. Therefore, the complete physics (or at least an important part of it) is given when the large group and the special representation of it are specified. In particular, by exploiting the properties of the Casimir operators it should be possible to calculate the mass spectrum contained in a given representation of the large group. The method was illustrated with the aid of two non relativistic examples where the exact solution of the Schrödinger equation is known, viz. the isotropic harmonic oscillator in n dimensions and the Kepler problem. For these two cases we were explicitly shown how the large groups which can be used for the purpose of calculating the energy spectrum look, how the Casimir operators of these groups should be manipulated, and which particular representation of these large groups should be used.

The large group turns out not to be unique. For the case of the harmonic oscillator two examples of groups both of which gave the correct energy spectrum were given (Sp(n,R) and SU(n,1)). Similar calculations were also given for relativistic theories and Budini showed us how this method could be used to calculate the mass spectrum of e.g. a Majorana theory.

Everybody realizes that there is a great difference between the problem of finding a large group for a given system with a known energy spectrum and the problem of finding THE GROUP in elementary particle physics as well as the special representation of it which gives the physically observed mass spectrum. The guiding principle for the invention of this large group (if it exists) is still to be found. Budini himself certainly realizes this point as well as - or better than - anyone else. However, this work is still in its beginning and we have to wait and see what comes out in the future.

Another subject which is an absolute "must" in any up to date conference about elementary particle physics is the current algebra. The lectures by Nauenberg about this subject during the first week have already covered by Pietschmann in his summary. During the second week we heard a lecture by Oakes about the application of current algebra to the calculation of certain branching ratios in the leptonic decays of K-mesons. Previous work in this field has indicated certain formal difficulties in the standard method of extrapolating the decay amplitudes off the mass shell to a point where the total four momentum of one (or more) of the π-mesons vanishes. This is a very important step in the argument when one wants to relate the amplitude for one decay to the amplitude for a similar decay but with one π-meson less in the final state. Oakes suggests that the formal difficulty noticed earlier has its origin in the assumption that all form factors have a very smooth momentum dependence. By explicitly introducing some terms with rather violent momentum dependence and suggested by certain Feynman diagrams he obtains a result which does not obviously suffer from the same difficulties as the previous calculations. In this way he obtains predictions about the decay rates for $K_{\ell 5}$ and $K_{\ell 4}$ decays which can, in principle, be compared with experiments. However, the decay rates are very low and the processes have not yet been observed.

In two lectures Joos has given a description of a very interesting phenomenological analysis of photo-production experiments. If I have understood him correctly the general idea is the following. He writes the electromagnetic current as a sum of terms each one having special symmetry properties under isotopic spin rotations etc. Next, he considers matrix elements of this operator between the vacuum and one particle, vector-meson states. As the vector mesons have definite isotopic spin quantum numbers etc., only one of the terms in the decomposition of the current will contribute to such a matrix element. Therefore, and as a definition, he calls each of these terms (apart from a numerical factor) "the field" for the vector particle under consideration. This field can then be used to evaluate the cross section in a reaction where such a vector meson appears in the initial state (e.g.). In a formal way one then gets a relation between the cross sections for an electromagnetic process and the same process where the photon is replaced by the vector meson. However, in reality one has to realize that the two matrix elements which enter in this way are evaluated at quite different points in momentum space (on different mass shells, e.g.). Therefore, they could, a priori, be quite different numbers. At this point the physically very important assumption is made that the matrix elements of the current (and fields) defined this way vary slowly over those parts of momentum space which are being considered. Such an argument leads e.g. to relations of the following kind

$$\sigma(\gamma + A \rightarrow B + C) = \text{const.} \times \sigma_{tr}(V + A \rightarrow B + C),$$

i.e. a relation between a photoproduction cross section and the corresponding cross section where the photon is replaced by a vector meson with transverse polarization. This result is equivalent with what is obtained in the so called "vector dominance model".

From the point of view of principle it must first of all be emphasized that the "fields" for the vector particles defined this way are not the usual fields used in elementary field theory. Nevertheless, they have some field like properties and if you want to call them fields no one can forbid you to do so. However, one should not expect that such fields fulfill all relations which one is used to in a normal field theory with canonical quantization. The important question here is not the names used but whether or not the smoothness assumption of the matrix elements of the "fields" is fulfilled. This is a question which cannot be decided from first principles. However, a comparison between the formula shown above as well as other relations of a similar nature and experiments indicates that the assumption is reasonably well satisfied for several cases. Therefore, this kind of method appears to be a possible way of analyzing photoproduction processes.

Among other subjects being covered during this last week I also have to mention a lecture by Thirring about the quantization of a scalar field in a de Sitter world. This is a problem which is perhaps not immediately related to any burning experimental issue but it is an amusing exercise in the mathematics of field theory. Actually Thirring considered a model world with one space and one time dimension and demonstrated that if this world has a de Sitter curvature one can still make a formalism with a consistent quantization of a free scalar field. At the end he stated, however, that some rather peculiar phenomena occur when interactions are introduced, and that a spontaneous creation of particles seems to take place. The actual rate of creation is

very small and, e.g., several orders of magnitude less than the rate of creation required in certain macroscopic models of the universe.
Also, Nauenberg gave us a review of the present status of CP-violating experiments. When the CP-violating $K_L \rightarrow 2\pi$ decay was discovered three years ago several models were proposed differing e.g. in the way in which the CP-violating amplitudes were introduced. Recent experiments have ruled out some of these models. In particular, the so called "superweak" model is by now definitely contradicted by the fact that the decay rate to a final state with two charged π-mesons is not the same as the decay rate to a final state with two neutral π -mesons.

Apart from these subjects we have also heard talks about radiative corrections to β-decay and about Coulomb fission. However, I am, for obvious reasons, unable to give you any opinion about the radiative corrections to β-decay which has not already been said by the lecturer. Also my knowledge about Coulomb fission is so incomplete, that I do not feel competent to review the lecture. Therefore, I will leave these two subjects out of this summary.

It only remains to thank the organizers of this meeting, in particular Professor Urban, for the very nice time they have given us all this year. I am sure I am speaking for everybody present if I say that we are all hoping to be able to come back to Schladming many times in the future.